Approaches to Social Inequality and Difference

Series Editors
Edvard Hviding, University of Bergen, Bergen, Norway
Synnøve Bendixsen, University of Bergen, Bergen, Norway

The book series contributes a wealth of new perspectives aiming to denaturalize ongoing social, economic and cultural trends such as the processes of 'crimigration' and racialization, fast-growing social-economic inequalities, depoliticization or technologization of policy, and simultaneously a politicization of difference. By treating naturalization simultaneously as a phenomenon in the world, and as a rudimentary analytical concept for further development and theoretical diversification, we identify a shared point of departure for all volumes in this series, in a search to analyze how difference is produced, governed and reconfigured in a rapidly changing world. By theorizing rich, globally comparative ethnographic materials on how racial/cultural/civilization differences are currently specified and naturalized, the series will throw new light on crucial links between differences, whether biologized and culturalized, and various forms of 'social inequality' that are produced in contemporary global social and political formations.

More information about this series at
https://link.springer.com/bookseries/14775

Marie Sandberg · Luca Rossi · Vasilis Galis ·
Martin Bak Jørgensen
Editors

Research Methodologies and Ethical Challenges in Digital Migration Studies

Caring For (Big) Data?

Editors
Marie Sandberg
Centre for Advanced Migration
Studies, the SAXO Institute
University of Copenhagen
Copenhagen, Denmark

Vasilis Galis
Center for Digital Welfare
IT University of Copenhagen
Copenhagen, Denmark

Luca Rossi
Digital Design
IT University of Copenhagen
Copenhagen, Denmark

Martin Bak Jørgensen
Department of Culture and Learning
Aalborg University
Aalborg, Denmark

Approaches to Social Inequality and Difference
ISBN 978-3-030-81225-6   ISBN 978-3-030-81226-3   (eBook)
https://doi.org/10.1007/978-3-030-81226-3

ACKNOWLEDGMENTS

This book is a result from the interdisciplinary research project DIGINAUTS Migrants' Digital Practices in/of the European Border Regime, funded by the VELUX Foundations 2018–2021.

THE VELUX FOUNDATIONS

VILLUM FONDEN ✖ VELUX FONDEN

PRAISE FOR *RESEARCH METHODOLOGIES AND ETHICAL CHALLENGES IN DIGITAL MIGRATION STUDIES*

"In a world where migrants and data are often treated with disrespect, this book offers a much-needed antidote to academic, political, technological and methodological carelessness. A must read for anyone who cares"
—Huub Dijstelbloem *is Professor of Philosophy of Science, Technology and Politics and Scientific Director of the Institute for Advanced Study of the University of Amsterdam, The Netherlands*

CONTENTS

NOTES ON CONTRIBUTORS

Martin Bak Jørgensen is Professor at DEMOS at the Department for Culture and Learning, Aalborg University, Denmark. He works within the fields of sociology, political sociology, and political science. He has published the books *Politics of Dissent* (Peter Lang, 2015; co-authored with Óscar García Agustín) and *Solidarity Without Borders: Gramscian perspectives on migration and civil society alliances* (Pluto Press, 2016) and *Solidarity and the 'Refugee Crisis' in Europe* (Palgrave, 2019) both co-authored with Óscar García Agustín). He is the Principal Investigator for the interdisciplinary project "DIGINAUTS: Migrants' digital practices in/of the European border regime".

Leandros Fischer has worked as a Postdoctoral Researcher for the DIGINAUTS project and was in charge of Subproject 2 (Hamburg and German–Danish borderland). His research interests include labour and migration, the re-negotiations of citizenship, as well as the welfare-migration nexus, and the role of the Left. His research has appeared in peer-reviewed edited volumes as well as journals, such as *Citizenship Studies*, *Mobilities*, and *Antipode*. He has taught at the universities of Aalborg, Marburg, and Cyprus (Nicosia).

Vasilis Galis is Associate Professor in the Technologies in Practice (TIP) group at the IT University of Copenhagen. Galis' research is interdisciplinary and it is impregnated by a strong epistemological solidarity with social movements. Galis has published on social movements, migration, and sociotechnical systems from a Science and Technology Studies (STS) perspective. He is co-Investigator for the interdisciplinary project "DIGINAUTS: Migrants' digital practices in/of the European border regime".

Ahmad Kamal is a Senior Lecturer in the Department of Cultural Sciences at Linnaeus University, Sweden, where he teaches in the digital humanities as well as library and information science. He joined the DIGINAUTS Project as a postdoc at the IT University of Copenhagen. His earlier research explored the circulation of contentious information in post-2011 Egypt and the roles of NGOs in promoting literacy in central India.

Koen Leurs is Assistant Professor in the Department of Media and Culture, Utrecht University, the Netherlands. He is the Chair of the European Communication Research and Education (ECREA) Diaspora, Migration, and the Media section. Using creative, digital, and participatory methodologies, Leurs seeks to engage in co-creative knowledge production on migration, gender, diaspora, youth culture, and digital technologies. Recently, he co-edited the *Handbook of Media and Migration* (Sage, 2020), and journal special issues including *"Migrant narratives"* for *the European Journal of Cultural Studies* (2020) and "Forced migration and digital connectivity in(to) Europe" for *Social Media + Society* (2018). He is currently a Fellow at the Netherlands Institute of Advanced Studies (NIAS), working on the monograph Digital Migration (forthcoming with Sage).

Anna Lundberg is Professor of Welfare Law at Linköping University and Associate Professor in Human Rights. Lundberg's research has appeared in, among others, *Critical Policy Studies*; *Human Rights in Practice*; *Refugee Survey Quarterly*; *International Journal of Law, Policy and the Family*; *Peace Review, Nordic Journal of Migration Research*; *Nordic Journal of Social Work*. Anna is one of the initiators to, and research leader of, "the Asylum Commission" (https://liu.se/en/research/asylum-commission).

Vasiliki Makrygianni has a background in Architecture Engineering and Urban Planning. She has worked as a researcher at the interdisciplinary project DIGINAUTS (ITU) (2018–2020) where she investigated the digital practices of migrants en route to Europe and on arrival sites.

Her primary research areas include critical urban theory and feminist methodologies while her current research interests focus on digital spaces and ICTs and feminist technoscience.

Nina Grønlykke Mollerup is Associate Professor at the Centre for Advanced Migration Studies and the Ethnology section, University of Copenhagen, Denmark. She is a Media Anthropologist with a specialisation in the Middle East. Her research focuses on the documentation of violence and conflict, and the use of media for knowledge-making and truth-making. She was a Postdoc in the interdisciplinary research project DIGINAUTS.

Anders Munk is Associate Professor at the University of Aalborg in Copenhagen, Director of the Techno-Anthropological Laboratory (TANTlab) and co-Founder of the Public Data Lab. He holds a D.Phil. in Geography from the University of Oxford and has worked as a visiting research fellow at the SciencesPo médialab. His research focusses on digital methods for controversy mapping and computational anthropology more broadly.

Luca Rossi is Associate Professor, at the IT University of Copenhagen, Denmark. He is member of the Networks Data and Society (NERDS) research group and of the Digital Platforms and Data research group. He also coordinates the Data Science & Society research lab. His interdisciplinary research tries to connect traditional sociological theory with computational approaches. Within this line of research, he has developed research in the context of online participation, online activism, political campaign, and election studies. He is Project Manager in the interdisciplinary project "DIGINAUTS: Migrants' digital practices in/of the European border regime".

Marie Sandberg is Associate Professor in Ethnology and Director of the Centre for Advanced Migration Studies (AMIS), the SAXO Institute, University of Copenhagen, Denmark. Her research focuses on everyday life Europeanisation, European borders, and migration practices. As a former joint Editor-in-Chief of *Ethnologia Europaea—Journal of European Ethnology* (2013–2020), a current member of the IMISCOE board of directors, as well as the elected President of SIEF—International Society for Ethnology and Folklore, she is actively engaged in Nordic and international research fields of ethnology and migration studies. She is co-Investigator for the interdisciplinary project "DIGINAUTS: Migrants' digital practices in/of the European border regime".

Kevin Smets is Assistant Professor at the Department of Communication Studies, Vrije Universiteit Brussel, and co-Director of ECHO: Research group on media, culture, & politics. He was trained as a Cultural Historian and obtained Ph.D. in Film Studies and Visual Culture (Antwerp, 2013). He currently leads the ERC Starting Grant project "Reel Borders" on how borders are imagined through film.

Ninna Nyberg Sørensen is a Senior Researcher at the Danish Institute for International Studies (DIIS), where she also heads the "Migration & Global Order" department. She has worked on international migration for most of her career and published widely on transnational migration, conflict, development, and gender. Her recent work explores high-risk migration (including human trafficking and forced return) as a consequence of increasingly rigid migration policies and border control practices. Much of her work is carried out in dialogue with scholars and institutions in the global South.

Laura Stielike is a Postdoctoral researcher in the research group "The Production of Knowledge on Migration" at the Institute for Migration Research and Intercultural Studies (IMIS) at the University of Osnabrück. She works on knowledge production in the field of big-data-based migration research and governance, the relations between migration and development, postcolonialism and intersectionality, and discourse and apparatus analysis.

Giacomo Toffano is a Ph.D. Fellow of the Research Council—Flanders (FWO) and a member of ECHO, the research group on media, culture, and politics at the Vrije Universiteit Brussel, Belgium. He obtained his MSc in Journalism and Media (2020—VUB), and his research now focuses on genre-hybridisations in migration narratives. In particular, Giacomo studies the production of hybrid media contents on migration that combine fiction with journalism, cartography, and data visualisation. He also holds an M.Sc. in Political Science—European Studies (Università degli Studi di Padova—2014) and has worked for five years as Accredited Parliamentary Assistant at the European Parliament.

LIST OF FIGURES

Caring for (Big) Data: An Introduction to Research Methodologies and Ethical Challenges in Digital Migration Studies

Marie Sandberg and Luca Rossi

INTRODUCTION—THE SCOPE OF THIS BOOK

Migration, historically, is a technologically supported process. However, the current migration influx into Europe is characterised by an elaborate use of digital technological applications. Nation-states and the EU border regime apply smart technologies to control and privilege the movements of some, while restricting and criminalising the movements of others (Hess and Kasparek 2017). On the other hand, irregularised migrants and networks of solidarity use Internet and Communication Technology

M. Sandberg (✉)
Centre for Advanced Migration Studies (AMIS), SAXO Institute, University of Copenhagen, Copenhagen, Denmark
e-mail: sandberg@hum.ku.dk

L. Rossi
Digital Design, IT University of Copenhagen, Copenhagen, Denmark
e-mail: lucr@itu.dk

© The Author(s) 2022
M. Sandberg et al. (eds.), *Research Methodologies and Ethical Challenges in Digital Migration Studies*, Approaches to Social Inequality and Difference, https://doi.org/10.1007/978-3-030-81226-3_1

(ICT) to facilitate passageways, thereby subtly reconfiguring how the digital platforms themselves function (Darling and Bauder 2019; Galis et al. 2016; Gillespie et al. 2016). Smartphones, for example, alleviate information precarity by providing access to networks of care as well as to in/formal work, while their meaning and uses vary depending on class, education, gender, and age (Walker et al. 2014; Wall et al. 2015, see also Vammen et al. 2021, 58). Digital technologies reshape not only every phase of the migration process itself—by providing new ways to access, share, and preserve relevant information—but also the activities of other actors, from solidarity networks to border control agencies. In doing so, digital technologies create a whole new set of challenges for migration studies: from data access to research ethics and privacy protection. When vulnerable and politicised groups like irregularised migrants constitute the primary research group, they face the risk of being (unintentionally) exploited and of unforeseen consequences based on their research participation (Pittaway et al. 2010). If issues of security, trust, and informed consent are already significant when researching migration (Zapata-Barrero and Yalaz 2020; Gillespie et al. 2016; Mackenzie et al. 2007), digital migration research only highlights those ethical challenges, adding further issues of privacy, (online) security, confidentiality, autonomy, and informed consent.

Regardless of the centrality played by technology in irregularised migrant trajectories and journeys, there is surprisingly little research that reflects on these new ethical and methodological challenges from a multidisciplinary perspective. Drawing on an interdisciplinary group of scholars that spans across critical border and migration studies, social media studies, anthropology of migration, and science and technology studies, this book offers an in-depth analysis of the most crucial methodological and ethical challenges in digital migration studies and reflects on ways to move this field forward. When digital technology becomes a lens and tool for shared decision-making and navigation among migrants, and at the same time an entrance for state authorities' surveillance and control, an update of our methodological approaches along with careful ethical considerations is urgently required.

This book therefore addresses methodological implications and ethical challenges when researching migrants' digital practices in the configuration of migration and borders. In this introductory chapter, and throughout the book, we use the term *irregularised migrant*.[1] While we apply the term *migrant* in a general and not juridical sense, we choose the adjective *irregularised* to stress the inability of migrants to travel through established and safe means and describe how they find themselves navigating through illegalised and often highly dangerous ways to safety; only too often with deadly consequences.

The computational turn within social science and digital humanities has proliferated new data formats and not least new questions for research (Boellstorff and Maurer 2015; Blok and Pedersen 2014). Whereas the so-called "big data" refer to data accessed on the basis of computational social science methods through API or data scrapings from social media platforms such as Facebook and Twitter, ethnographic materials are generated on the basis of qualitative research methods and ethnographic fieldwork, including in-situ engagements like participating observations, face-to-face conversations, in-depth semi-structured interviews, and online "netnography" (Kozinets 2019). Yet, as discussed in this book, how can differences between apparently disparate data formats be conceptualised and how do we remain sensitive towards the fact that the computational tools and digital platforms themselves guide options for asking questions about the material collected? How can "big social data" and qualitative and/or ethnographic materials be brought into closer dialogue and which ethical implications should be considered? How can we aim for more in-depth analysis of migrants' digital traces, in ethically sound ways, when access to context knowledge is limited, if not absent? Perhaps the great divide between quantifiable data sets and qualitative insights requires rethinking. As suggested by Munk (2019), digital traces are at one and the same time quantitative and qualitative, since traces such as likes and shares, can be counted, while they also contain rich text, such as comments and profile data.

Whereas digital methods have grown into an established field that cross-fertilises media studies, STS, computer science, and information design (Rogers 2013, see Munk in this book), the intersections between migration and border studies and computational methods and digital ethnography are less developed. For migration and border studies, it is therefore of particular interest to discuss the challenges in drawing on digital data, which comprise computational big data on the one hand and ethnographic materials on the other. Crafting, relying on, and combining these data and material types in new ways make questions regarding data access, data interpretation, privacy protection, and research ethics generally even more pertinent. For instance, how can informed consent be ensured in online digital fora or social media platforms, and if not, how should ethical research be conducted? What relations of reciprocity are possible and feasible when working on virtual, and often very interchangeable, temporary digital platforms? How can we ensure and promote migrants' capacity for autonomy when pursuing digital migration research? The pertinence of these questions appears to be even stronger since the online presence of research participants is only brief and meant to be untraceable and unidentifiable.

This book discusses digital migration research methodology and ethics when conducting and combining qualitative and ethnographic fieldwork accompanied by digital data analysis. Central aspects concern both the generating of data (e.g. multiple platforms, different API, data accessibility) as well as data analysis (inconsistent data, missing data, context-dependent data). Of specific concern are the aspects of digital migration researchers accessing digital platforms used by migrants, who are subject to precarious and insecure life circumstances, lack recognised papers, and are in danger of being rejected and deported. How does the digital migration researcher ensure that the scrutinisation of online activity does not jeopardise migrants' lives and safety?

Crucially, the methodological considerations concern an ongoing discussion and reflection on the kind of knowledge digital migration researchers produce, and how to avoid compromising research participants' safety before, during, and after research is conducted. Engaging in ethically sound relationships between researchers and migrant research participants through the principles of integrity, respect, autonomy, and justice have long since been the ethos in the context of migration research (Hynes 2003; Voutira and Doná 2007). Further calls for moving beyond minimal standards of "doing no harm" to research participants in vulnerable positions have been made in order to establish more viable relationships, including reciprocal benefits to participating migrants or migrant communities (Mackenzie et al. 2007). Yet, for digital migration studies, the question remains how we can preserve and strengthen similar types of careful ethical and methodological approaches when dealing with migrants' digital data.

In this book we argue that working with digital technologies and large-scale data sets in relation to ethnographic studies of digital migration practices and trajectories requires new modes of *caring for (big) data*. Besides the already mentioned issues of taking proper care of research participants' privacy, autonomy, and security, this also spans carefully establishing analytically sustainable environments for the respective data sets (see also Sandberg, Mollerup and Rossi in this book).

As a notion underpinning the contributions in this book, we propose the *notion of care* in the context of ethical and methodological considerations for digital migration studies, through inspiration from the work of empirical philosopher and ANT researcher Annemarie Mol. She coins the notion "the logic of care" (Mol 2008) to highlight *care work* as an inclusive and open-ended process integral to daily life involving a range of heterogeneous actors and relations (see also Mol et al. 2010). Even though Mol's notion of care is developed in another context and its aim differs to that of this book, as Mol presents a critique of the neoliberalised Dutch health-care system, we find her thoughts inspiring for our purpose of furthering our discussion of ethical issues in digital migration research.

Mol encourages researchers to engage in the problem of care from the beginning of any research engagement, which can likewise help to identify the questions migration and border scholars need to ask when dealing with ethical and methodological research implications. Importantly, following Mol's concept of care, care work does not rely solely on individuals but is rather distributed as a matter of concern for a set of heterogeneous and sociomaterial actors cooperating in specific situational settings. In order to de-individualise issues of ethics and methodological practices by reaching out not only to the single researcher, but also to influence the ethos of the research collective, we suggest turning our attention to the logic of care. We thus take Mol's call to pursue and nurture the logic of care as a point of departure for highlighting and nurturing the care work as a prism for how this book's contributors deal with digital data in migration research.

With the notion of (better) caring research as a point of departure, the book presents reflections on research design and methods that move beyond state-of-the-art methodologies for discussing how to combine or merge quantitative and qualitative methods with the prospect of transgressing boundaries between online and offline data. Ultimately, this will facilitate more viable research on the complex, cross-platform nature of migrants' information and communication technology (ICT) use.

In this introduction, we will first recapture some main characteristics of ethical and methodological considerations within migration studies and discuss how this field has sought to move beyond the "do no harm" approach (Stierl 2020). We will then present the emerging field of digital migration research in order to pinpoint the specificities of the ethical and methodological considerations required within this field of research. Rather than starting from ground zero, we propose to learn from and nurture the already established research ethics developed within migration studies and related fields.

Second, we turn our attention to the issue of big data and how to care for those data in ethically viable ways. Here we discuss the remaining question concerning how we can preserve the same type of careful approach when dealing with digital data as when conducting ethnographic and qualitative research with vulnerable groups such as migrants and people living in the context of insecure and violent circumstances.

Third, we outline the chapters of this book and how they contribute new avenues and ways for furthering the ethical dimension of digital migration research. In conclusion, we highlight how these chapters, rather than proposing any quick fixes or set solutions, offer alternative viewpoints and stimulate critical thinking on the part of border authorities engaging with migrants' digital practices, as well as migration researchers, in order to strengthen and promote ethically rigorous research.

Ethical and Methodology Challenges in Migration Studies

Migration studies as a field characterised by its interdisciplinarity has a long track record of using a multiplicity of research methodologies and approaches. This quest for multiplicity is based on the maxim that complex phenomena, such as migrational issues, call for insights, perspectives, and contributions from several disciplines. Because of the moving field and wide geographical distribution of migration studies, research questions can be difficult to answer with a singular method. However, as argued by Voutira and Doná (2007, 166), some certain characteristics still unite research on refugees and migration as a field of study, namely, a genuine interest in pursuing bottom-up perspectives to further migrants' points of view, along with a blurred line distinguishing between advocacy and scholarship (ibid., 167). Likewise, more state-centred perspectives (e.g., in international relations, law, and economics) in migration studies tend to include critical perspectives on migration policy and border regimes (ibid.).

A further joint characteristic can be added to the field of migration studies, namely, a distinct preoccupation with research ethics engaging the principles of integrity, respect, autonomy, and justice (Hynes 2003;

Voutira and Doná 2007; Zapata-Barrero and Yalaz 2020). Witnessing one's research (unintentionally) harming the subject of research, the community of the research subject in question, or being used politically to further certain agendas, is undoubtedly the worst-case scenario for any migration scholar. However, as argued by Mackenzie et al. (2007), migration research has since long been in dire need of moving beyond minimal standards of "doing no harm" when researching participants in vulnerable positions and subsequently establishing more viable relationships between researchers and research participants, including the enabling of reciprocal benefits to participating migrants or migrant communities (Mackenzie et al. 2007, 300, see also Stierl 2020). Mackenzie et al. highlight the need for recognising and promoting migrants' agency and autonomy in terms of capabilities and rights (ibid., 302), a call also reflected in several chapters of this book, which draw on inspiration from the Autonomy of Migration (AoM) approach.

In digital migration studies, as we argue, the question remains how we can preserve the same kind of careful ethical and methodological approaches when dealing with migrants' digital practices and digital data. In the following, we discuss how to move this mode of caring for our research participants in migration research ethics into the context of digital data.

Digital Migration Research—Past and Future Methodological Challenges in an Emerging Field

While still relatively young, digital migration research is quickly consolidating into a fully fledged academic field. It already satisfies most of the criteria that we usually adopt when defining an academic field of research: dedicated special issues (Leurs and Smets 2018, 1), international conferences (e.g., connectingeuropeproject.eu), as well as a certain level of internal reflection (Leurs and Prabhakar 2018; Andersson 2019, and many chapters in this book). In their introduction to the special issue for the international journal *Social Media + Society*, Leurs and Smets (2018, 1–16) focussed on the context surrounding the emerging field by asking a set of questions including the following, which we find extremely relevant for our purpose: Where are the field and focus of digital migration studies? And where is the human in digital migration? Discussing these questions seems particularly relevant for understanding how a careful ethical and methodological approach has emerged and can be further developed in digital migration research.

Migrants were using media technologies, such as letters from friends who had already migrated; news and video content from "across the border" (Mai 2001, 95) first; and digital media later, long before digital migration studies acknowledged that digital traces were a viable data source. Although Appadurai stressed the connection between global migration and digital technologies already in 1996, it took twenty years for digital migration studies to attract global attention as a necessary and timely approach for understanding contemporary migration. What happened during those twenty years is of interest when understanding the promises and the expectations that accompanied the early days of digital migration studies. For the first decade of the twenty-first century, digital data were already used to study online communities, which offered researchers unprecedented access to diasporic communities around the world (Komito and Bates 2009, 232). From this perspective, the combination of digital data and migration studies emerged within the context defined by digital ethnography (Hine 2008; Markham 2005) where online communities (of migrants) and their (digital) practices were the object of research. Later, the focus on what was deemed to be possible to study with digital data changed. The combined effect of social media and digital traces (Giglietto et al. 2012, 145; Venturini and Latour 2010) as well as growing social awareness of the use of large amounts of digital data to analyse social phenomena (Kitchin 2014) created the perfect background for a paradigm shift. Instead of studying specific communities that researchers had to access through ethnographic principles, the digital traces that migrants were leaving behind in the form of GPS coordinates, social media posts, likes, or shares, contributed to the idea that it was possible to study migrations through "data only" without the need to engage with the producers of those data: the migrants. This built on the parallel emergence of data from social media platforms and it was considered a comprehensive—and sometimes preferable—research option for social scientists and digital researchers alike (ibid., 1; Felt 2016, 1). Despite several attempts to call for a critical reflection on the epistemological consequences of this data revolution (Kitchin 2014, 1) and the emergence of several empirical limitations (Tufekci 2014, 505), the new perception of data reached migration research in combination with the historical events that once more afforded migrations, asylum seekers, and migration-related policies centre stage in global discussions. The so-called "European refugee crisis", which continued for most of the second decade of the century, provided the "perfect" societal context for larger-than-ever use of digital data in migration studies and for the parallel datafication of migrations (Leurs and Smets 2018, 4). Migrants and asylum seekers became represented by their digital data in a variety of different communication artefacts, policy reports, and academic research, and the ways in which this happened, including the narratives underlying this process, are far from inconsequential.

This process of an accelerated craving for big, digital migration data was facilitated by the strong anchoring of digital migration studies in the field of (digital) media and communication research. Several of the main theoretical and methodological approaches are either native to the fields of media studies or have been used in the context of media studies for decades (see, for example, Canidatu et al. 2019, 36). The perceived social relevance connected with the "European refugee crisis" in Europe, the societal predisposition towards the ongoing data revolution, and the availability of a set of research methods and practices from compatible academic fields, made the first 15 years of digital migration research an addition to the existing field of migration studies rather than an extension of it. As Yalaz and Zapata-Barrero (2018, 14) point out in their work covering 15 years of qualitative migration studies, between 2000 and 2016, the overall number of articles published in migration studies doubled, though this increase was not produced by an explosion in the qualitative approach to migration studies, which remained a stable quota over the years. The explosion was due to the growing production of quantitative and digital methods and approaches to migration studies that gained new relevance during this period. Beside the enthusiasm for a new and promising set of research methods, it should be noted how something quite unique happened when digital methods reached the area of migration studies. Entire research domains that in a pre-digital methods scenario would have previously required a considerable amount of contextual knowledge and in-situ relations became available, just a few clicks away, to a much larger group of scholars.

Within this process, digital migration studies, rather than building its ethical stand and approaches by expanding the more careful approach defined by qualitative migration research, often adopted a media-centric (Smets et al. 2019) approach in which ethical concerns focused more on the data than on the subjects behind it. This does not mean in any way that existing digital migration research lacks ethics, but that it has, so far, not fully delivered on Mackenzie et al.'s (2007) idea of moving beyond the minimal ethical standard. How is it possible for research based on the digital traces left by migrants to enable reciprocal benefits for migrants or migrant communities? (Ibid., 300). How do we recognise autonomy and agency when the migration process is, partially or entirely, datafied? How do we care for all the data that allows us to research migrations? While there are no easy answers to these questions, we think that many of the chapters in this collection show that these are very pertinent questions. As we argue, rethinking the way in which data is understood in digital migration studies in more careful ways can result in a stronger connection between issues and theoretical apparatus.

It Is Big Data—Who Cares?

Following the ideological enthusiasm for the ongoing data revolution, data has emerged as the technological solution for any type of border control (Bigo 2014, 209–225; Broeders 2007, 71–92). Today more than ever, human mobility is represented, studied, and governed through big data. Large data sets of biometric data promise to protect the (smart) borders by combining efficiency with safety (Sontowski 2018, 2730–2746) and when data is not available, ad-hoc initiatives are launched to fill the gaps.[2] It is fair to say that governments' and other international actors' interest in migrants' digital data has never been greater, and this is especially true of data that is perceived to be useful for policing smart borders (ibid.) or preventing allegedly illegal immigration (Latonero and Kift 2018).

Within this scenario, it should be clear that the unintended and unforeseen consequences of migrants' digital data collected for research purposes can be nefarious, and that the legal privacy-oriented procedures in place in many research institutions (e.g., GDPR compliance) might not even achieve an adequate state of "doing no harm". For these reasons, we see that more care is required regarding how we approach digital migration data for research. We see the logic of care being adopted in research projects as a continuous open-ended process (Mol, 2008) that could be summarised in two interrelated steps: firstly, before the data collection, and secondly, during and beyond the active time of research. As a first step, approaching migrants' digital data "with care" means pursuing a more critical approach to the use of big data in migration research where the data is not an unquestionable proxy for social activity. From this perspective, the relations between the social practices behind the data production are fully understood, and their links with the issue of migration research are clearly conceptualised to avoid unnecessary and potentially harmful data collection. This form of care builds on the idea of curation that Munk suggests (2019, 164) as a way of bridging the quali-quantitative divide. Munk defines curation as "Critically reappropriating (and thus manually curating) onlife traces to speak on behalf of certain phenomena or address certain questions" (ibid.). Caring for big data used in migration research points in a similar direction. While it is certainly possible to imagine a non-manual form of curation, the key element is the re-appropriation of the digital traces within the theoretical design of the research and making methodological decisions on that basis.

The second step of a careful approach to migrants' digital data is to create an analytically viable and sustainable environment for the research data. Research practices rooted in digital methods are often expected to share the data sets that have been used for the research efforts. Replicability of the research results as well as the possibility for further research are common arguments used to support this request (Weller and Kinder-Kurlanda 2016, 166). While the practice of data sharing is well established in the context of clinical data (Bull et al. 2015, 225–238) and some practices have been adopted for sharing social media data (Benton et al. 2017, 94–102), there seems to be very little guidance when it comes to migrants' digital traces. Data sharing, as well as post-research data storage, needs to be balanced against the interests of all the actors involved, bearing in mind that data value and potential harm caused by data are not stable over time. This needs to balance apparently conflicting aspects: on the one side, as Weller and Kinder-Kurlanda note (2016, 170), the reproducibility value of digital traces deteriorates over time while, on the other side, the risk of personal harm exists even when a considerable period of time has passed, as a growing body of legal instruments have acknowledged (e.g., in the—still limited—implementation of the so-called "right to be forgotten" specified in Article 17 of GDPR).

Digital migration studies have huge potential to provide insights into some of the most relevant issues of our time. Nevertheless, the very same characteristics that make the approach powerful and have contributed to its rapid growth as an academic research field can easily represent a risk for the subjects involved. This calls for a more critical, reflective, in other words more careful, approach to big data in the context of migration studies.

OUTLINE OF THE BOOK

The chapters in this book reflect an interdisciplinary theoretical framework that draws on methodologies from critical border and migration studies (cf. Casas-Cortés et al. 2015), social media studies (cf. Rodríguez et al. 2014; Croeser 2014), anthropology (cf. Ingold 2018; Strathern 1991/2004), and science and technology studies (STS) (cf. Dijstelbloem and Meijer 2011). The methods used include policy analysis, qualitative approaches entailing non-participant observation, ethnographic interviews, and media device tours as well as data-scraping techniques for analysing social media data. This multiplicity of research methods is a deliberate editorial choice, as we believe that analysing the dynamics and consequences of borders, mobilities, and technologies requires a multi-faceted methodological toolbox.

A theoretical premise of the book's research insights is that borders are not fixed geographical entities but a set of complex practices in a constant state of becoming, and that technology transforms not only migration but also forms of solidarity with migrants. The analyses presented in this volume therefore not only include migrants' use of ICT, but also solidarity networks and groups facilitating refugee reception. This, in turn, requires careful ethical considerations when working with data gained from migrants' stories as well as digital imprints from solidarity networks facilitating irregularised border crossings.

The idea for the current book was developed as part of the interdisciplinary research initiative called "DIGINAUTS: Migrants' digital practices in/of the European border regime" that began in 2018, funded by the Velux Foundation Denmark. The DIGINAUTS project argues that migrants' uses of technology not only challenges our usual ways of thinking about migration but also subtly reconfigures the functioning of these technologies themselves.

The methodological implications of working with digital data in migration studies and thus ideas relating to the further focus of this book were extensively discussed during a methods workshop at the IMISCOE conference in Malmö in June 2019 as well as a workshop that took place in Copenhagen in October 2019, hosted by the Ethos lab at the IT University Copenhagen, both organised by the DIGINAUTS project. During these activities, it was clear that a growing number of migration researchers are aware of the possibilities offered by digital data but are facing methodological and ethical challenges. We have invited contributions from researchers with whom we have worked closely during the book project as well as researchers working with topics highly relevant for this book. Together, these contributions represent state-of-the art within critical migration studies as well as digital migration studies and social media studies.

This introductory chapter is followed by two parts, each with a set of chapters. Part I highlights "Digital and Qualitative Data Dynamics" whereas Part II scrutinises in detail "Ethical Challenges in Digital Migration Research and Beyond". The book concludes with a third "Comments" part, in which three researchers, each with distinguished research expertise in their respective fields of migration studies and digital research, offer concluding reflections and comments spanning the contributions in this book.

The first part "Digital and Qualitative Data Dynamics" contains four chapters that present the many facets of digital data in the context of dealing with migration. The part contains specific proposals to conduct research with migrants' digital data, both directly and through derivative products, as well as overviews describing the field of digital migration studies in its complexity. In Chapter 2, "Migrant Digital Space: Building an Incomplete Map to Navigate Public Online Migration", the authors Vasiliki Makrygianni, Ahmad Kamal, Luca Rossi, and Vasilis Galis discuss the challenges encountered while sampling online data from a largely unknown population and especially so from "minor actors" such as the digital spaces set up by migrants. They reflect on these challenges by introducing the concept of Migrant Digital Space as an online (and offline) arena where information, knowledge, communication, advocacy, and representation of migrants are enacted by leveraging contemporary digital technologies. From this perspective, migrant digital space is inherently unstable, and its definition is an integral part of any research on migrants' digital practices.

In Chapter 3, "Contrapuntal Connectedness: Analysing Relations Between Social Media Data and Ethnography in Digital Migration Studies", Marie Sandberg, Nina Grønlykke Mollerup, and Luca Rossi explore the potentials of combining ethnography and "big" social media data in analysing fieldwork carried out with Syrian refugees and solidarians in the Danish–Swedish borderlands 2018–2019, as well as data collected during 2011–2018 from 200 public Facebook pages run by solidarity organisations, NGOs, and informal refugee welcome and solidarity networks. The authors suggest that the relationship between the types of research material can be conceived as contrapuntal, which means that the material types are recognised as different but fundamentally inter-connected. Inspired by Tim Ingold (2018), lines of counterpoint (known from musical theory when different musical lines are played simultaneously, while being at once independent and related) are translated into human movements, which carry on alongside one another, not as a summation of parts but as the correspondence of its particulars. This contrapuntal connectedness is explored and further qualified with the aim of identifying potentials and further questions for digital migration research.

Chapter 4, "Migration Trail: Exploring the Interplay Between Data Visualisation, Cartography and Fiction" by Giacomo Toffano and Kevin Smets, departs from a case study on Migration Trail, an online interactive platform, and discusses the potential in techniques that visualise migration. Data visualisations, in this perspective, present migration scholars with a new set of problems and ethical challenges. What narratives of migration emerge from the way data are visualised, and who bears responsibility for those narratives? The authors apply a mixed-method approach that includes both multimodal and discourse analysis to understand and scrutinise the interaction of textual, audio, visual, and spatial elements of communication in Migration Trail.

The final chapter in this part, Chapter 5, "Migration Multiple? Big Data, Knowledge Practices and the Governability of Migration" by Laura Stielike, explores the production of knowledge on migration at the interface between migration research built on big data and governance. Applying discourse analysis to research papers based on big data, the chapter carves out characteristic features of such migration studies. In her work, Stielike highlights the risk of big-data-based migration research connecting with pre-existing narratives about migration that present it as an object of government.

The second part of the book "Ethical Challenges in Digital Migration Research and Beyond" contains three chapters that all zoom in on the ethical challenges faced during digital migration research. Chapter 6, "Impossible Research? Ethical Challenges in the (Digital) Study of Deportable Populations Within the European Border Regime", continues along similar lines. The authors, Leandros Fischer and Martin Bak Jørgensen, discuss and reflect on the ethical challenges faced when conducting ethnographic research and online ethnography among groups facing deportation. They consider the implications of doing or not doing such research and discuss whether this is "impossible research", due to national authorities being provided with access to data that migrants would prefer remained less visible. Migration researchers, the authors argue, should adhere to ethical principles of working with vulnerable groups such as migrants scheduled for deportation, without compromising their sense of agency. Taking its cue from a "militant research approach" along with the "autonomy of migration" (AoM) perspective, the chapter argues for reflexive and contextualised ethics that aim to promote solidarity and social change.

Chapter 7, "The Redundant Researcher: Fieldwork, Solidarity and Migration" by Vasilis Galis, does not offer solutions to this ethical challenge but reveals a set of critical, productive self-reflections on the author's own research practices. On the basis of fieldwork conducted on the islands of Lesvos and Chios during the winter of 2019, the chapter asks: What is it like to conduct academic research on a phenomenon that is polluted by vested political interests, personal strategies, ideological loyalties, propaganda, and hazards? Why is this fieldwork relevant and for whom? In order to answer these questions, the chapter proposes four principles for an emancipatory migration research paradigm to ensure that the research conducted promotes migrants' agency, addresses concerns relevant to migrants themselves, supports migrant struggles, and ensures the safety and integrity of migrants. Whereas the two preceding chapters discuss, in different ways, how to "do no harm" or how to use research to stipulate and empower migrant struggles, Chapter 8, "Emotional Introspection: The Politics and Challenges of Contemporary Migration Research" by Ninna Nyberg Sørensen, discusses an often-overlooked question: How to do no harm to ourselves, as researchers, when doing migration research in demanding and stressful situations often embedded in ethically challenging contexts? Based on long-term observations and experience within the field, combined with a set of recently conducted interviews with migration researchers, Sørensen discusses the institutional cultures and structures in migration research in the context of stricter migration policy and practice. Sørensen argues that we need to attend to the emotional aspects of conducting fieldwork in complex, increasingly more insecure and challenging situations. Along with making the emotional implications more explicit, emotional introspection is therefore called for, before, during, and after research.

The concluding three shorter commentaries address both of the main issues of this book: Big data—research methodologies and ethical challenges in digital migration studies as well as reflecting on the contributions of this book. Koen Leurs, an expert in digital migration studies, reflects in his comment "On Data and Care in a Migration Context"; Anders Kristian Munk, who is very well versed in the digital humanities and mapping of this controversial field, argues that we should consider "Caring as Critical Proximity: A Call for Toolmaking Digital Migration Studies"; and Anna Lundberg, a migration research scholar with a keen interest in welfare law and academic activism, asks: "What Should We Do as Intellectual Activists? A Comment on the Ethico-political in Knowledge Production".

Conclusion

In conclusion, this book provides a unique contribution to the emerging field of digital migration research by bridging insights from critical migration and border research, anthropology of migration, feminist theory, science and technology studies (STS) with social media and communications research within digital humanities. These research approaches have in common that the exceptionality and irregularity of categories such as "refugee" and "migrant" are critically and self-reflexively assessed. The authors thus argue that it is essential to carefully reflect on researchers' own positioning as being part of the research challenges they seek to address. By devoting special attention to the links between digital research methodologies and ethics in migration studies, the chapters cover innovative approaches that intersect digital social media studies, critical border and migration studies, and ethnography, and aim to contribute to ongoing and emerging debates on research ethics in digital migration research and the complex entanglements of migration with technology. The following chapters should stimulate a much-needed critical reflection on ethical and methodological issues in digital migration research. As we have argued, researching migrants' digital practices in the configuration of migration and borders calls for new modes of *caring for (big) data*. Besides taking proper care of research participants' privacy, autonomy, and security, this also spans carefully establishing analytically sustainable environments for the respective data sets, as outlined here. Finally, we aspire this book to be used by an interdisciplinary readership consisting of migration scholars and students alike, and that by stimulating further methodological discussion in our fields, it will enable collective reflection related to the ethics of digital migration.

NOTES

1. For a discussion regarding the term "refugee" as a "categorical anomaly", see Voutira and Doná (2007, 163).
2. Very interesting examples of this are the "Filling data gaps" initiatives launched by the UNHCR joint data center on Forced Displacement: https://www.jointdatacenter.org/what-we-do/#filling-data-gaps.

BIBLIOGRAPHY

Andersson, Kerstin B. 2019. "Digital Diasporas: An Overview of the Research Areas of Migration and New Media through Narrative Literature Review." *Human Technology* 15 (2): 142–180.

Appadurai, Arjun. 1999. "Globalization and the Research Imagination." *International Social Science Journal* 51 (160): 229–238.

Benton, Adrian, Glen Coppersmith, and Mark Dredze. 2017. "Ethical research protocols for social media health research." In *Proceedings of the First ACL Workshop on Ethics in Natural Language Processing*, pp. 94–102.

Bigo, Didier. 2014. "The (In) Securitization Practices of the Three Universes of EU Border Control: Military/Navy–Border Guards/Police–Database Analysts." *Security Dialogue* 45 (3): 209–225.

Blok, Anders, and Morten Axel Pedersen. 2014. "Complementary Social Science? Quali-Quantitative Experiments in a Big Data World." *Big Data & Society* 1 (2): 2053951714543908.

Boellstorff, Tom, and Bill Maurer. 2015. "Introduction." In *Data, Now Bigger and Better!*, edited by Boellstorff Tom and Maurer Bill. Chicago: Prickly Paradigm Press.

Broeders, Dennis. 2007. "The New Digital Borders of Europe: EU Databases and the Surveillance of Irregular Migrants." *International Sociology* 22 (1): 71–92.

Bull, Susan, Nia Roberts, and Michael Parker. 2015. "Views of ethical best practices in sharing individual-level data from medical and public health research: A systematic scoping review." *Journal of Empirical Research on Human Research Ethics* 10 (3): 225–238.

Candidatu, Laura, Koen Leurs, and Sandra Ponzanesi. 2019. "Digital Diasporas: Beyond the Buzzword: Toward a Relational Understanding of Mobility and Connectivity." *The Handbook of Diasporas, Media, and Culture*, 31–47. Hoboken: Wiley.

Casas-Cortes, Maribel, Sebastian Cobarrubias, Nicholas De Genova, Lenda Garelli, Giorgio Grappi, Charles Heller, sabine Hess, Bernd Kasparek, Sandro Mezzadra, Brett Neilson, Irene Peano, Lorenzo Pezzani, John Pickles, Federico Rahola, Lisa Riedner, Stephan Scheel and Martina Tazzioli. 2015.

"New Keywords: Migration and Borders." *Cultural Studies* 29 (1): 55–87. https://doi.org/10.1080/09502386.2014.891630.

Croeser, Sky. 2014. "Changing Facebook's Architecture." *An Education in Facebook*, 185–195.

Darling, Jonathan, and Harald Bauder, eds. 2019. *Sanctuary Cities and Urban Struggles: Rescaling Migration, Citizenship, and Rights*. Manchester University Press.

Dijstelbloem, Huub, and Albert Jacob Meijer, eds. 2011. *Migration and the New Technological Borders of Europe*. London: Palgrave Macmillan.

Ehn, Billy, Orvar Löfgren, and Richard Wilk. 2015. *Exploring Everyday Life: Strategies for Ethnography and Cultural Analysis*. Lanham: Rowman & Littlefield.

Felt, Mylynn. 2016. "Social Media and the Social Sciences: How Researchers Employ Big Data Analytics." *Big Data & Society* 3 (1): 2053951716645828.

Galis, Vasilis, Spyros Tzokas, and Aristotle Tympas. 2016. "Bodies Folded in Migrant Crypts: Dis/Ability and the Material Culture of Border-Crossing." *Societies* 6 (2): 10.

Giglietto, Fabio, Luca Rossi, and Davide Bennato. 2012. "The Open Laboratory: Limits and Possibilities of Using Facebook, Twitter, and YouTube as a Research Data Source." *Journal of Technology in Human Services* 30 (3–4): 145–159.

Gillespie, Marie, Lawrence Ampofo, Margaret Cheesman, Becky Faith, Evgenia Iliadou, Ali Issa, Souad Osseiran, and Dimitris Skleparis. 2016. "Mapping Refugee Media Journeys." *Smartphones and Social Media Networks*. Research report. The Open University. http://www.open.ac.uk/ccig/sites/www.open.ac.uk.ccig/files/Mapping%20Refugee%20Media%20Journeys%2016%20May%20FIN%20MG_0.pdf.

Hess, Sabine, and Bernd Kasparek. 2017. "Under Control? Or Border (as) Conflict: Reflections on the European Border Regime". In *Perspectives on the European Border Regime: Mobilization, Contestation, and the Role of Civil Society*, edited by Ove Sutter and Eva Yokhama. *Social Inclusion* 5, 58–68.

Hine, Christine. 2008. "Virtual Ethnography: Modes, Varieties, Affordances." *The Sage Handbook of Online Research Methods*, 257–270.

Hynes, Tricia. 2003. "New Issues in Refugee Research." *The Issue of 'Trust' or 'Mistrust'in Research with Refugees: Choices, Caveats and Considerations for Researchers*. Geneva: Evaluation and Policy Analysis Unit, The United Nations Refugee Agency.

Ingold, Tim. 2018. *Anthropology: Why It Matters*. Cambridge: Wiley.

Kitchin, Rob. 2014. "Big Data, New Epistemologies and Paradigm Shifts." *Big Data & Society* 1 (1): 2053951714528481.

Komito, Lee, and Jessica Bates. 2009. "Virtually Local: Social Media and Community among Polish Nationals in Dublin." In *Aslib Proceedings*. Bingley: Emerald Group Publishing Limited.

Kozinets, Robert V. 2019. *Netnography: The Essential Guide to Qualitative Social Media Research*. London: Sage.

Latonero, Mark, and Paula Kift. 2018. "On Digital Passages and Borders: Refugees and the New Infrastructure for Movement and Control." *Social Media + Society* 4 (1): 2056305118764432.

Leurs, Koen, and Kevin Smets. 2018. "Five Questions for Digital Migration Studies: Learning from Digital Connectivity and Forced Migration in (to) Europe." *Social Media+ Society* 4 (1): 2056305118764425.

Leurs, Koen, and Madhuri Prabhakar. 2018. "Doing Digital Migration Studies: Methodological Considerations for an Emerging Research Focus." *Qualitative Research in European Migration Studies*, 247–266. Basel: Springer.

Mackenzie, Catriona, Christopher McDowell, and Eileen Pittaway. 2007. "Beyond 'Do No Harm: The Challenge of Constructing Ethical Relationships in Refugee Research." *Journal of Refugee Studies* 20 (2): 299–319.

Mai, Nicola. 2001. "The Role of Italian Television in Albanian Migration to Italy." *Media and Migration: Constructions of Mobility and Difference*, edited by Russell King and Nancy Wood, 95–109. New York: Routledge.

Markham, Annette N. 2005. "The Methods, Politics, and Ethics of Representation in Online Ethnography." In *The Sage Handbook of Qualitative Research*. Thousand Oaks, CA: Sage.

Mol, Annemarie. 2008. *The Logic of Care: Health and the Problem of Patient Choice*. Abingdon: Routledge.

Mol, Annemarie, Ingunn Moser, and Jeanette Pols. 2010. "Care: Putting Practice into Theory." In *Care in Practice. On Tinkering in Clinics, Homes and Farms*, edited by Annemarie Mol, Ingunn Moser, and Jeanette Pols, 7–26. Bielefeld: Transcript Verlag.

Munk, Anders Kristian. 2019. "Four Styles of Quali-Quantitative Analysis: Making Sense of the New Nordic Food Movement on the Web." *Nordicom Review* 40 (1): 159–176.

Pittaway, Eileen, Linda Bartolomei, and Richard Hugman. 2010. "'Stop Stealing Our Stories': The Ethics of Research with Vulnerable Groups." *Journal of Human Rights Practice* 2 (2): 229–251.

Rodriguez, Manuel Gomez, Krishna Gummadi, and Bernhard Schoelkopf. 2014. "Quantifying Information Overload in Social Media and Its Impact on Social Contagions." In *Proceedings of the International AAAI Conference on Web and Social Media*, vol. 8, no. 1.

Rogers, Richard. 2013. *Digital Methods*. Cambridge, MA: MIT press.

Smets, Kevin, Koen Leurs, Myria Georgiou, Saskia Witteborn, and Radhika Gajjala, eds. 2019. *The Sage Handbook of Media and Migration*. London: Sage.

Sontowski, Simon. 2018. "Speed, Timing and Duration: Contested Temporalities, Techno-political Controversies and the Emergence of the EU's Smart Border." *Journal of Ethnic and Migration Studies* 44 (16): 2730–2746.

Stierl, Maurice. 2020. "Do No Harm? The Impact of Policy on Migration Scholarship." In *Environment and Planning C: Politics and Space*, October 2020. https://doi.org/10.1177/2399654420965567.

Strathern, Marilyn. 1991/2004: *Partial Connections*. Updated edition 2004: Washington, DC: Altamira Press, Rowman & Littlefield.

Thoreau, Henry David. 2016. "Walking." In *The Making of the American Essay*, edited by John D'Agata, 167–195. Minneapolis: Graywolf Press.

Tufekci, Zeynep. 2014. "Big Questions for Social Media Big DATA: Representativeness, Validity and Other Methodological Pitfalls." In *Proceedings of the International AAAI Conference on Web and Social Media*, vol. 8, no. 1.

Vammen, Ida M. S., Sine Plambech, Ahlam Chemlali, and Ninna Nyberg Sørensen. 2021. "Does Information Save Migrants' Lives? Knowledge and Needs of West African Migrants en Route to Europe". *DIIS Report*, 2021, no. 1: 1–65. Copenhagen.

Venturini, Tommaso, and Bruno Latour. 2010. "The Social Fabric: Digital Traces and Quali-quantitative Methods." In *Proceedings of future en seine* 87–101.

Voutira, Eftihia, and Giorgia Doná. 2007. "Introduction. Refugee Research Methodologies: Consolidation and Transformation of a Field." *Journal of Refugee Studies* 20 (2): 163–171.

Walker, Rea, Lee Koh, Dennis Wollersheim, and Pranee Liamputtong. 2014. "Social Connectedness and Mobile Phone Use Among Refugee Women in Australia." *Health & Social Care in the Community* 23 (3): 325–336.

Wall, Melissa, Madeline Otis Campbell, and Dana Janbek. 2015. "Syrian Refugees and Information Precarity." *New Media & Society*, 1–15. Published online before print July 2, 2015. https://doi.org/10.1177/146144481559 1967.

Weller, Katrin, and Katharina E. Kinder-Kurlanda. 2016. "A Manifesto for Data Sharing in Social Media Research." In *Proceedings of the 8th ACM Conference on Web Science*, 166–172.

Yalaz, Evren, and Ricard Zapata-Barrero. 2018. "Mapping the Qualitative Migration Research in Europe: An Exploratory Analysis." In *Qualitative Research in European Migration Studies*, 9–31. Cham: Springer.

Zapata-Barrero, Richard, and Yalaz Evren. 2020. "Qualitative Migration Research Ethics: A Roadmap for Migration Scholars." *Qualitative Research Journal* 20 (3): 269–279. https://doi.org/10.1108/QRJ-02-2020-0013.

PART I

Digital and Qualitative Data Dynamics

Migrant Digital Space: Building an Incomplete Map to Navigate Public Online Migration

Vasiliki Makrygianni, Ahmad Kamal, Luca Rossi, and Vasilis Galis

INTRODUCTION

In 2018, the interdisciplinary research project "DIGINAUTS—migrants' digital practices in/of the European border regime" set off to explore the information practices of migrants and solidarity networks across specific European regions along contemporary migration routes (Greece, Germany, Denmark, and Sweden). Central to this effort was the role of

V. Makrygianni (✉)
Copenhagen, Denmark

A. Kamal
Department of Cultural Sciences, Linnaeus University, Växjö, Sweden
e-mail: ahmad.kamal@lnu.se

L. Rossi
IT University of Copenhagen, Copenhagen, Denmark
e-mail: lucr@itu.dk

M. Sandberg et al. (eds.), *Research Methodologies and Ethical Challenges in Digital Migration Studies*, Approaches to Social Inequality and Difference, https://doi.org/10.1007/978-3-030-81226-3_2

25

information and types of communication technology (ICT) in circumventing and challenging the European border regime (see also Sandberg
and Rossi in the Introductory chapter of this volume). To address the
implicated role of ICT in contemporary migration, the research team
applied a mixed-methods approach that would bring together offline
ethnographic fieldwork with large-scale online data analysis. While the
analytical potential of this approach, and the methodological and epistemic frameworks underlying it, are explored by Sandberg, Mollerup, and
Rossi in Chapter 3 of this volume, it is useful to explain the initial context
in which the DIGINAUTS project evolved. The idea was to produce
a map of the digital resources and spaces that constitute a considerable part of the informational background of contemporary migration.
For instance, a Syrian refugee's account of her experiences in Lesvos
(Lesbos), Greece, needs to be contextualised within the public content
from migrant-oriented pages and websites based in Greece that provide
part of the set of informational resources drives the journey. Within this
perspective, DIGINAUTS had neither a solely online focus and nor did
the project aim complementing the two data sources but at understanding
the relationship between online information and offline actions. This is
exemplified by the contrapuntal approach, developed and presented in
Chapter 3.

In order to make the integration between large-scale digital data and
ethnographic observations possible, DIGINAUTS aimed at mapping the
space where the digital encounter of migrants and solidarity workers
takes place—which we refer to as migrant digital space (MDS). In digital
migration studies, such endeavours are rare (Gualda and Rebollo 2016),
as most studies of migrants online rely on interview-based studies (Dekker
et al. 2018), smaller-scale online studies, or topic- or group-specific large-
scale data collections (Kok and Rogers 2017). When compared to the
wealth of studies on digital diasporas (Laguerre 2010), there are few
studies of the digital networks of in-transit migrants; yet even fewer
studies are conducted mainly online, see, for instance, the data-orientated
research of Kok and Rogers (2017) and Sánchez-Querubín and Rogers
(2018). Most scholars investigating migrants' adaptation of digital

V. Galis
Center for Digital Welfare, IT University of
Copenhagen, Copenhagen, Denmark
e-mail: vgal@itu.dk

space have relied primarily on first-person interviews (Dekker et al. 2018; Gillespie 2018; Leurs 2014), ethnographies (Grzymala-Kazlowska and Phillimore 2018), or surveys (Merisalo and Jauhiainen 2020), to identify relevant online sites, user preferences, and the benefits afforded.

The DIGINAUTS researchers set out to analyse DMS through publicly available user-generated content before research teams operating independently undertook the main part of the ethnographic work. While the sub-team in charge of the digital data collection researched initially at the various sites (Germany, Greece, and Øresund region encompassing the Danish–Swedish borderlands) to identify valuable seeds (relevant online resources and actors) for the data collection, the tools, and the techniques needed to be established in advance.

In this chapter, we develop an analytical framework for understanding the ways in which migrants and digital spaces are intertwined, and we investigate (digital) spaces that allow for circumventing borders, solidarity practices, and for shaping migration. This materialises into what we define as Migrant Digital Space (MDS). We position MDS within the existing research of migrants and digital technologies, we briefly report on the steps and guidelines we adopted to collect the data, and we present a qualitative overview of MDS, as we have defined it, highlighting some specific characteristics and possibilities.

Migrants' Digital Traces

The study of migrants' engagement with ICT has long been of acute interest to researchers from various fields, such as migration studies, media and communication studies, critical border studies, anthropology, and science and technology studies (see, for instance, the works of Alencar 2018; Aouragh 2011; Brinkerhoff 2009; Diminescu 2008; Leurs 2014, 2015; Leurs and Smets 2018; Madianou and Miller 2012; Siapera 2014). Recent works (Drüeke et al. 2019; Latonero and Kift 2018, also Gillespie 2018), as well as the special-issue *Forced Migrants and Digital Connectivity in Social Media and Society* journal (2018) and the volume *The Sage Handbook of Media and Migration* (Smets et al. 2019) provide a complete overview of the complexity of the issue. Several studies and research projects were developed during the 2015 European migration crisis, and the number of publications on the topic is growing. Research projects such as "WhyWePost" (Miller et al. 2016), "CON-NECTINGEUROPE: Digital Crossings in Europe: Gender, Diaspora and

Belonging" (Ponzanesi 2016), "Resilient Communities, Resilient Cities? Digital makings of the city of refuge" (Georgiou 2013) as well as the DIGINAUTS research project have highlighted the impact of ICT on the everyday life of migrants.

This impressive quantity of work covers many different directions of the possible relation between migrant populations and ICT and has evolved considerably over the years to account for new dynamics, new technologies, and new concerns. Rather than offering a complete overview of the whole field of research, in the following we organise some of the existing research based on how these recent contributions approached the idea of migrant space within digital migration studies and how that has affected how MDS was developed within DIGINAUTS.

ICT has often been studied in the guise of tools able to provide access to valuable informational resources. The nature of these resources as well as "when" the resources are thought to be used in the migration processes varies considerably: researchers (Dekker et al. 2018) highlighted the role of ICT in providing access to valuable information before the journey, while others (Alencar 2019) have shown how migrants who are already settled in the country of destination, as well as solidarians, adopt various kinds of ICT to offer sought-after resources leading to the emergence of "transglocalised" networks (Kok and Rogers 2017). As they argue, the particular territorial arrangement and engagement of the digital are what form a transglocalised network where local networked formations exist alongside national and transnational formations, each operating with awareness of the other yet acting separately.

As our case study shows, the thematic of online activity vary and depend on the different means, the distinct digital subjects, and the various geolocations. For instance, while in many cases researchers have focussed on practical information both to reach the country of destination as well as to facilitate a resettlement process, ICT can also provide migrants with valuable social and emotional capital (Marino 2015). Also, it is notable how practices of care develop in the digital sphere (Leurs 2019; Alinejad and Ponzanesi 2020) among individuals and communities and produce *digital spaces of care* (Makrygianni and Galis, forthcoming). In this context, it is interesting to observe how, while ICT is generally described in terms of potential access to information, there is a parallel interest focussing on how trust in specific information is constructed (Dekker et al. 2018; Mollerup and Sandberg, forthcoming). As Gillespie et al. (2016) explain, there is a paradox between the growth of available information and the lack of timely, relevant, and reliable information that makes the migrants' decision of which informational resource should be trusted more complex.

This growing body of research depicts the social, digital, and physical space where the migration process takes place as extremely complex, populated by various actors and supported by different types of "infrastructures" (Dekker et al. 2018). Media and communication studies have shown how migrants create digital spaces of homemaking, and becoming (Georgiou 2013; Leurs 2015; Madianou and Miller 2012; Witteborn 2014; Xie and Witteborn 2019). Smets (2018) discusses the idea of "mobile homes" and Almenara-Niebla (2020) the "digital home-camps" to expand the field of research beyond media studies. Latonero and Kift (2018) bring the notion of "digital passages" to grasp both the temporal condition and the architectural constraints of the migrants' movement within this space. Similarly, Gillespie et al. (2018) draw on the work of Smets (2018) and Roseneau (2003) to discuss the concepts of "digital passages" and "distant proximities" while Leurs (2015) also builds on the passage in order to address issues of navigation in digital space. In addition, research on place-making practices that relate to digital activities has started gaining ground (Bork-Hüffer 2016; Lim et al. 2016; Polson 2015; Witteborn 2012, 2015). Recently, Smets et al. (2019) have addressed issues of space in a chapter of the already highlighted volume *The Sage Handbook of Media and Migration* dedicated to the spatial dimension. There, most of the authors (Alencar; Alevizou; Costa and Wang; Xi and Witteborn) elaborate on a more relational approach to space and pay special attention to reterritorialising processes and place-making practices.

These various ways of defining the space where migration takes place are often quite diverse in terms of the actors that "populate" and produce this space, in terms of the size of the space itself and in terms of the challenges that it poses. We claim that some aspects of this diversity can be explained by looking at the various methods used to "map" the space. After all, both the data we collect about any space as well as the research methods and processes (related, for instance, to different scales or timelines) will, unavoidably, define our conceptualisations and representations of it.

Within this perspective, we chose to adopt a data-centric lens to organise the existing research. Data-centric means that we organised the existing research about migrants and refugees according to their data-collection strategies: ethnographic approaches, selected groups of sources, and large-scale data collections.

Research with a strong qualitative orientation, usually conducted either through multi-sited ethnography (Charmarkeh 2013) or in-depth interviews with selected subjects (Dekker et al. 2018), defines the migrant space both through the specific migrant population and the specific actors they include in their research design. The size and breadth are defined by the phenomenological nature of both the analysis and, mostly, the data-collection practices. Within this approach, the definition of the borders of the migrant space is determined by the actual experiences of the subject, either directly reported or mediated through the ethnographic practice or the researcher.

An alternative approach involves research aimed at mapping the digital experience within a specific geographical context (Kaun and Uldam 2018) or within a specific type of actor (Dessewffy and Nagy 2016). In this case, the actors are usually pre-selected and then their digital activity is tracked or monitored. The size and breadth of the space are defined by the set of actors that are included in it. To limit the possible problem deriving from this approach, the set of actors that are studied can be obtained or complemented with computational methods based on crawling of digital resources (Kok and Rogers 2017).

A third approach is what can be defined as large-scale data collection. In this case, actors are usually identified using the affordances of a pre-existing digital platform or social media service. This could have different practical implementation depending on the digital data the research is actually working with: it could be searching for specific keywords on Facebook or following specific hashtags on Instagram or twitter. For example, Siapera and colleagues (2018) used a set of "refugee-related" hashtags to identify a data set of 7,500,000 tweets. In this way, they were able to represent a discussion space that is largely co-opted by mainstream entities (politicians, media, NGOs) and frames (humanitarian vs. far right) and would probably be invisible through an alternative type of data collection.

Those three strategies have different scopes, goals, and ambitions. While this diversity is largely a natural consequence of the variety of research methods and preferred data, we claim that a careful design of a mixed-methods approach allows for constant alterations of the scale of the research that minimises the unavoidable blind-spots of digital data collection. Therefore we suggest both a macroscopic level of analysis—a "birds-eye view" of the data—as well as a smaller-scale in-depth overview of the (digital) everyday practices of the digital subjects.

Configuring Migrant Digital Space

This section problematises and conceptualises migrant digital space that is configured by migrants' online activity before the journey, en route, and when settling. By conceptualising MDS as a space shaped by practices, we aim to (a) understand the various digital-place-making practices, (b) investigate the relation between human mobility and digital, and (c) map migrants' spatial footprints in the digital sphere. We follow a relational approach to space (Lefebvre 1991; Massey 2005; Harvey 2006), according to which, space is not considered as a life container but as a derivative of social relations and interactions. Therefore, we understand MDS as an outcome of social relations and practices with material and intangible characteristics.

With the concept of MDS, we conceptualise an online (and offline) arena where information, knowledge, communication, advocacy, and representation of migrants is enacted by leveraging contemporary digital technologies. This space is formed by (a) digital subjects (accounts, pages, hashtags, channels) touching on (b) migrant-related topics (such as discussions on migration routes; language lessons; football conversations; university enrolment; job seeking) through conversations across (c) various digital platforms. MDS is thus first defined by this triple multiplicity: multiple actors, multiple topics, and multiple platforms. While previous studies have positioned physical spaces as either "a binary or an opposition to the perceived virtuality of the emerging web" (see Shah 2019), in our approach, we highlight the interconnection and interdependence of analogue and digital spatialities. Migrants' online activities are formed from a simultaneous use of software (which requires time and communication skills) and hardware, i.e., devices, antennas, cables, and satellites. Moreover, MDS represents simultaneously a counterpoint to and an extension of migrants' everyday life in physical spaces (camps, borders, city centres, streets, plazas, parks, routes, trains, neighbourhoods, libraries, cafés, etc.) where they enact their agency and constraints, whether with respect to mobility, information seeking, social interactions, or entertainment, and so forth. Thus, MDS is synthesised by internet-based platforms, digital subjects, and elements of physical/material space.

In terms of form and function, MDS is a space for commoning that incorporates characteristics of a public space such as diversity, heterogeneity, and contradictions. Information sharing, facilitating access to goods and infrastructures and various solidarity acts thrive beyond commodification practices and constitute MDS as a space of commons rather than a space of commodities (An Architektur 2010). While we describe MDS as a space for commoning, it should be noted that the very same digital space spans several contradicting dynamics taking place: access to the MDS raises questions about the accessibility of the digital space (Khorshed and Imran 2015, 344), which emphasises pre-existing inequalities, as well as a growing issue of trust in the actors and content populating the digital space (Borkert et al. 2018, 1; Gillespie et al. 2018, 1). Within the new context, the individual ability to navigate these uncertainties as well as the technological divide that accompanies it become the main elements that define membership of the space rather than race, ethnicity, gender, or physical space characteristics and biases. Primarily, it is a space that enables encounters and conflicts among analogue and digital subjects. As this space for practices emerges from subjects, topics, and platforms and has a clear connection with the physical and material space, it should appear clear that we can hardly imagine dealing with a single, unique, MDS. On the contrary, we need to embrace the idea that MDS has a spatial dimension that necessarily results in multiple, partially overlapping, Migrant Digital Spaces since subjects, topics, and platforms combine differently for the different geographical contexts. The network of relevant resources, topics, and actors will, while partially overlapping for the European border regime, also varies due to local specificities.

The digital resources we have identified would thus carry relevant information differentiated according to different temporalities and geolocations. For example, as we will show more analytically further on in our data sets, the pages with a geolocation in Greece focus mainly on primary needs, in Germany on information on settling down, in Sweden on asylum seeking, and in Denmark on spaces for encounters and networking. These needs and topics differentiate especially during periods of crisis, such as the extreme weather conditions of winter 2015 in Greece (which we designate the "winterisation" period).

THE PRACTICE OF MAPPING MDS

As mapping an unknown space presents a series of methodological challenges, we had to employ a set of navigation tools. Based on the conceptual description outlined in the previous paragraph, we considered the migrant digital space as an unknowable and unstable set of digital resources available to migrants before, during, and after their journeys. We considered it unknowable because it is defined by an unknown number of entities that are, in many cases, hard to identify in a binary distinction between relevant and non-relevant entities. As soon as we move beyond the official organisations or larger NGOs, the digital space offering resources for migrants is characterised by small entities, often with an unclear status, which may highlight the specific issue only for a limited time period (e.g., in the midst of large humanitarian crises). We decided to adopt a migrant-centric perspective when defining what could constitute a "useful resource". This means that the criteria for a resource to be included in our mapping activity was its ability to provide information that could have been valued as relevant by the migrant population, rather than using any top-down criteria of relevance or authority. We soon realised that often the relevant information found in a large variety of digital spaces: from the above-mentioned digital presence of large solidarity organisations to small (often tiny) groups of solidarians or migrants—already living in one of the destination countries—is provided by people who, at a particular moment in time, decided to step up and help. These heterogeneous resources are clearly scattered across the digital world on various platforms: from Facebook to websites, from WhatsApp groups to telegram channels. Each digital platform is clearly accompanied by its own set of technical affordances and user expectations. Moreover, as a consequence of the wide-ranging actors and motivations, the migrant digital space is also unstable. The abundance of minor and informal actors that, thanks to the opportunities offered by contemporary social media platforms, entered the digital space to provide relevant information for migrants also produced high instability in the space. Mapping, within an unstable digital environment, has a very clear—and often short—temporal dimension, it is an act of representation of spacetime. Online spaces are often created and become inactive (or are left abandoned) within months, information becomes old and outdated, and the general space is constantly reshaped. This instability lies behind what Gillespie and colleagues (2018) describe as a "Paradox": a growth in available information that corresponds to a growth in uncertainty due to the inadequacy or unsustainable nature of this information. Thus, mapping is an incomplete process that exposes the incoherence and fragmentations of migrants' digital spatialities. As a research practice, it is not a taxonomy or an ordering process of migrants' everyday digital practices but an incomplete action, "a simultaneity of unfinished stories", as Massey (2003) describes space itself.

Given these premises, the effort to map the Migrant Digital Space had to be anchored to four axes of research:

(a) A user-centric perspective

When selecting the actors that would populate the migrants' digital space, we decided to focus on migrants looking for information. We did that using a combination of keywords from specific searches both in the local languages (German, Greek, Danish, and Swedish) as well as in community languages (Arabic) and English to identify valuable online resources. We complemented those with pre-existing information that we identified in 12 pilot interviews conducted in Greece during the summer of 2016. Moreover, along with the online data gathering (spring 2018) we initiated our qualitative research and located four informants in the Øresund region (Copenhagen), six informants in Germany (Hamburg), and three in Greece (Athens) (all interviewees were migrants who arrived in Europe in about 2015). We asked them to guide us in their digital spaces and paths while suggesting the most and less popular and trustworthy pages. Locating our digital actors and our interviewees was a result of online and offline snowballing.

(b) Ethical data collection

From the very beginning of the project, we decided that we would work only with publicly available data. This led to the exclusion not only of private mobile data (e.g., WhatsApp group conversations) but also of data that could have been perceived as private but that could have been accessible for research purposes (e.g., closed Facebook groups). While we adopted a migrant-centric perspective for snowballing sources, we opted for removing any identifiable information on the migrants from our data even when this meant being unable to trace the information-seeking activity of individual migrants. While we recognise that Facebook, as well as other platforms, (1) is centralised and the flow of information is vastly controlled and biased (see, for example, van Dijck 2009; Ho 2020), and (2) does not always secure a safe space for all (see, for example, cyber hate cases), the simultaneous presence and interaction of millions of people, regardless of the existing contradictions, suggest that it is perceived, at least partially, as a possible space for debate. We argue that such spaces constitute spaces of othering, spaces of healing, care, and emancipation for the migratory populations.

(c) Temporarily limited

As mentioned before, MDS is characterised by its ever-changing nature. While hypothesising that ongoing data collection is undoubtedly fascinating, practical reasons, as well as the restriction imposed by Facebook on data access (Bruns 2019), forced us to define a firm end for the data collection (24/09/2018). While the option of collecting past data gave us the opportunity to collect longitudinal data, the data should still be approached as inherently unstable since an unknown quantity of data could have already been deleted from the platform previous to our data collection.

(d) Alteration of scales

The mixed-method approach led us to simultaneously conduct research on various scales to conceptualise and represent MDS. Thus, we considered testimonies of individuals (that suggested popular pages according to personal criteria) as well as snow-balling our online research to come up with our data set. As feminist scholars note (Massey 2005; Smith 1987; Yuval Davis 2007), the alteration of scales (which entails more of a qualitative than a quantitative differentiation) examines practices at a macroscopic level and on a molecular basis and deconstructs the (digital) common sense. By focussing on migrants' everyday digital practices, we shed light on the various cultural geopolitical and social inequalities embedded by bordering practices. As we will show further on, the alteration of scales is a constant revisiting of migrant's everyday practices that reveals minor scale tactics of resistance against major scale institutional strategies of repression. Brenner (2001, 608) indicates that "the establishment and reorganization of scalar hierarchies creates geographies and choreographies of inclusion/exclusion and domination/subordination which empower some actors, alliances and organizations at the expense of others, according to criteria such as class, gender, race/ethnicity and nationality".

In our case, migrants' minor digital acts deconstruct hegemonic (institutional) narratives that reproduce the nation-state rationale and uncover places of discipline and power that spread from migrant's bodies to transnational territories. For instance, focussing on data coming from (less populated/less popular) LGBTQ+ pages (following a suggestion from some of our LGBTQ+ interviewees) revealed a spectrum of large-scale institutional bordering practices imposed on migrant's bodies due to their gender or sexual orientation.

The combination of these guidelines led to the definition of the data set we used to describe the MDS. On a practical level, constructing the digital data set developed according to the following three steps:

1. The research team identified an initial set of digital resources (448 public Facebook pages) that fulfilled the above-detailed criteria and corresponded to the specific geographical focus of the project.
2. The research team manually coded each page according to a set of criteria of interest for the research project. These were: the full names of the page, the main language used on the page, the geographical area of interest of the page (e.g., a German page could support search and rescue operations in the Aegean sea), the country of the organisation behind the page, the organisational level of the organisation behind the page (institutional, semi-institutional, non-institutional), and if the page focussed on LGBTQI issues.
3. Using the—now defunct—public Facebook API, we collected all the content publicly available on the pages from the creation of the page until the date of the data collection (September 2018). Given the shutdown of the Facebook API during the data collection, not all public pages that were initially identified (448) have actually been fully collected.

The final data set was composed of 200 Facebook pages that were then manually coded with additional information such as the type of actor behind each Facebook page, the physical location of the actor, the date when the page was created as well as the language (or languages) used. At the same time, all the content (posts, comments, and reactions) publicly available on the pages was downloaded using Facebook's API. This produced a final data set totalling 200 pages, 84,359 posts, and 2,254,923 comments, produced between 20/12/2010 and 24/09/2018.

In the following sections we show how the data collected in the MDS can be used to investigate the intersection of digital resources with migrants' everyday lives. This should not be understood as an in-depth analysis of the issue but more as a demonstration of the research that the mapping of the MDS makes possible. It should be acknowledged that this research can also be investigated with different data or strategies but, we claim, a throughout mapping of the relevant migrant digital space provides several benefits either because of the types of actors that can be included in the research, because of the longer longitudinal perspective that can be adopted, or because of a facilitate comparative perspective.

Migrants' Everyday Digital Places

There is a strong interdependence between migrants' everyday practices and digital space. Following the discussions on the notion of place that is formed by peoples' lived experiences (Massey 2005), digital places derive from the everyday life experiences and practices of migrants. Along the same line, thinkers from the field of feminist studies, sciences of space and social sciences (Smith 1987; Lefebvre 1977/2014; de Certeau 1980/2013; Massey 1994, 2005) have pinpointed the importance of everyday life in the (trans)formation of space. De Certeau in "Practices of everyday life" 1980/2013 investigated "routine practices", such as walking, talking, reading, and cooking, and he found creative resistance to these "arts of doing" of ordinary people. In a similar way, migrants, as minor actors (Margetts et al. 2016), perform various digital practices (digital arts of doing) such as group chats, video calls, microblogs, emails, online games, online music, and online shopping, posting, commenting on social media, sharing, (dis)liking, microblogging, and so on. Such digital arts of doing reveal a plurality of themes, relevant actors, and digital spaces that create migrants' digital places. While migrants' everyday lives take place both in physical and digital locations, several of their primal needs are fulfilled when forming their digital spaces of commoning. Their digital routine practices develop mainly around information sharing on the migration status and on the journey.

If we start looking for these everyday dimensions in the data set we collected, we can observe how the MDS is filled by a multiplicity of everyday problems and challenges that migrants address through digitally enabled commoning. However, we recognise that since we never laid eyes on private groups and certain platforms (due to ethical restrictions), there are limitations in our understanding of the notion of everyday in digital space. Among the many possible examples, we will briefly hint at a few that stood out in our reading of the data and constitute potential avenues for further analysis in digital migration studies.

The Evolution of Everyday Life Throughout the Journey

Figure 2.1 shows how most of the 64 non-institutional entries located in (various places in) Greece involved the distribution of food (e.g., No Border Kitchen). Fifteen pages developed around news and information about travelling and raised awareness of migrant issues (with one specifically aimed at "provid[ing] help and support from around the world with phone top-ups for refugees and displaced people". Another fairly common feature were pages with information on legal issues, pages about housing collectives (such as the "City Plaza", or the "Notara housing project"), education services (page of the "NO BORDER school"), and a page to support and raise money for medicines, food, water, etc. Less common were pages directly addressing LGBTQ+ solidarity or job

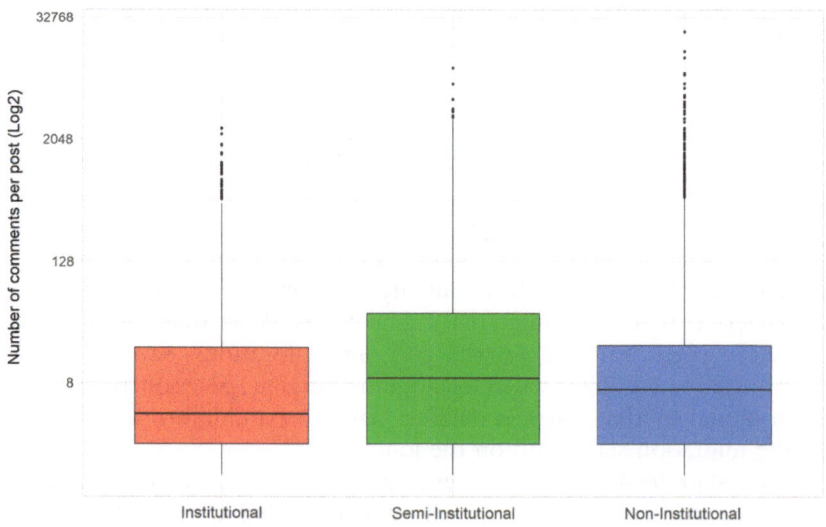

Fig. 2.1 Types of actors. This image is used with permission of the authors of this chapter [Rightsholders]

seeking (the page "Work opportunities for refugees in Greece" proudly proclaimed to be "the only one in Greece with this thematic").

The issues at the centre of the Facebook page activity change as we focus our attention on different geographical contexts.

Of the 125 pages located in Germany, two relate to LGBTQ+ solidarity, while the vast majority are about sharing information on refugee issues. We also found a page that acted as a Hamburg city guide (Yalla: "*We give you tips for Hamburg. We discover places, activities and groups in our city*"), and another offered "assistance for reunification" (Assistance for Syrians reuniting their families). Of the 83 pages located in Sweden, most involved information on asylum processes and one was specifically orientated towards LGBTQ+, once again providing information on asylum processes. Of the 40 entries in Denmark, most concerned networking and information on everyday life. One page was dedicated to practical issues (# HjælpEnFlygtning, which, as they mention, "*is the place where you can offer refugees a home, a job/an internship, language teaching, friendship/networking, or leisure activities*"), another offered legal advice (Jurarådgivning), and one was for LGBTQ+ asylum seekers (LGBTQ+ ASYLUM).

Everyday Vulnerability

Scholars such as Shah (2019) stressed the relation between the digital place forming practices with the critical discourse on vulnerability and the agency of bodies in the face of accelerated digitisation. Our data set contains various entries that involve the so-called "vulnerable subjects". According to some of our informants (involved with the "Lesvos Lgbtiq+ Refugee Solidarity" page), such pages act as empowerment tools, as they open the discussion on vulnerability and provide affect and compassion and sometimes alternatives. As mentioned above, we traced some LGBTQ+ solidarity groups in all three geolocations. In particular, in Greece we found three related pages: "Lesvos Lgbtiq+ Refugee Solidarity", "Lgbtqi+ Refugees" in Greece and "Eclipse". From previous field research, we found that these pages also generated physical meetings among the group members and various offline events of solidarity in analogue space. We also located one page from Germany dedicated to issues of bodily ability ("The National Association of the Deaf North Rhine-Westphalia, which 'supports deaf refugees from their arrival until they have settled in'"). Another German page was dedicated to parenting and guardianship (named "AKINDA", it "supports young refugees and minors who arrived in Berlin without parents").

Non-institutional Is More Trustworthy?

As mentioned in our mapping practices, we grouped the digital actors of our initial set of identified pages according to their institutional or non-institutional affiliation (of 448 entries, 317 are listed as non-institutional—see Fig. 2.1). This differentiation (between institutional, semi-, or non-institutional) highlights the different standpoints of the actors behind the pages and reveals the power geometries (Massey 1999) of migrant digital space. Such hierarchical division of the data can be used to make visible different strategies and tactics on behalf of the various actors in this spatial battle of "repression and expression" (De Certeau 1984). While potentially enacted also by "ordinary people", strategies are associated with hegemonic regulations and disciplinary mechanisms. In a migration framework, such regulations and mechanisms are to control both transnational mobility as well as the materialities of everydayness in migrants' settlements. Tactics, on the other hand, derive from daily practices of "ordinary people" that are not implied in certain institutional borders. As pointed out by De Certeau (1984, 117) and Yilmaz (2013, 68), these tactics are performed by those who are not included in the power group creating the power apparatus, such as workers, migrants, and subaltern ethnic minorities and tend to erode power mechanisms. Such digital tactics lack a permanent position and evoke distortions of (institutional) strategies of power, such as bordering and racialised practices. Applying this distinction to the data we collected, it means to understand how and if the institutional actors were providing informational resources aligned with the hegemonic regulations and if, on the opposite side, non-institutional actors were offering what we could define as tactical information.

When analysing our data set, we observed the differences between state and institutional strategies and the migrants' tactics on both the online and offline battlegrounds. Simply looking at the average number of comments received by the posts on the various types of pages shows how institutional pages, while having the largest numbers of followers, receive a smaller number of comments. From our field research, we know that mistrust in institutions led to augmented digital activity of non-institutional actors. Although "system avoidance" (Latonero and Kift 2018) is a practice of defence, the minor actors of our data set eventually took control of their own environment and restored a sense of identity, attachment, and belonging to their places of reference (whether these are spaces formed for shorter or longer periods). For instance, information regarding legal advice was circulated mainly by non-institutional actors (such as the Danish page for free legal advice, "Jurarådgivning" formed by law students at the University of Copenhagen), or the various pages in the German and Swedish digital spheres that come from the non-institutional listing and involve information for asylum seekers. Such processes of digital place-making that generate places of trust involve "processes of participation which include a conception of civic agency and the conditioning of space—affective, material, social, mediated" as Alevizou (2019) points out. These tactics of non-institutional actors seeking trustworthy online places differ according to geolocation. For instance, in the Greek case, it seems a large number of pages related to primal needs (such as housing and food) emerged especially around the year of 2015. Drawing from the data set but also from our informants' testimonies, the popularity of such non-institutional pages is not only due to mistrust towards the state authorities but also resulted from the state's failure to cover migrants' basic needs during the long summer of migration (2015).

Temporal Evolution of MDS

A final note on the challenges and possibilities of this time of approach to data collection should focus on the issue of temporal evolution of digital data. As noted above, MDS is understood as a time-sensitive entity, given the rapidly changing international and domestic context and the instability of the user population. The ephemerality of digital subjects within MDS is also a consequence of the unstable conditions faced by many of its actors and participants. For instance, in our case, this temporality was evident during the four months since between the two rounds of data collection (June and September 2018), already approximately 5% of the accounts had been discontinued. In addition, the acceleration of online information sharing fosters a compressed space that breaks traditional barriers. In this sense, we are witnessing the construction of various spatio-temporalities within the digital realm. Considering the different events during the migration crisis, we noticed very different temporal dynamics, depending on if the pages were of institutional, semi-institutional, or non-institutional actors. While very few institutional pages became active during the time covered by our data collection, we observed a sharp rise in non-institutional pages created between 2014 and 2017. This period included: (1) The "long summer of migration", which refers to the increase in migrants arriving in Europe during the summer of 2015, (2) The enactment of the EU–Turkey deal in March 2016, which resulted in around 50,000 migrants being "trapped" on Greek territory for several months, and (3) The so-called "winterisation period" (a term introduced by international NGOs referring to harsh weather conditions that exacerbated the suffering of thousands migrants in that region of Greece (see UNHCR 2015; Papataxiarchis 2016). During this period, many primal needs of migrants in Greece were fulfilled not by institutional actors but by non-institutional volunteers and activists. At this stage, several online and offline networks were activated as a reflexive response to the multiple crises. It is interesting to observe how the activity on semi-institutional pages peaked later than the non-institutional pages, suggesting that informal groups with little or no formal structure reacted quicker online to offline events than their formal counterparts.

However porous and unstable it may seem, MDS still facilitates analogue place attachments. Digital technologies reterritorialise specific activities and engagements. Brun describes the reterritorialising process of as "the way in which displaced people and local people establish new, or rather expand networks and cultural practices that define new spaces for daily life" (2001, 23). Within this context, migrants are articulating connections to various places and various actors at various moments in time. This makes MDS a temporarily stable space, kept in existence by the engagement and reterritorialising activity of its actors.

CONCLUSIONS

The overall methodological goal of the DIGINAUTS project was to apply a mixed-methods approach to bringing together offline ethnographic fieldwork with large-scale online data analysis of in-transit migrants, a rare endeavour in the literature. This was not merely for reaching analytical complementarity in mixing big and thick data sources (Bornakke and Due 2018). We aimed for a hybrid space where everyday experiences are entangled with a large pool of digital data, traces, information, and so on. We coined the concept of Migrant Digital Space to conceptualise this hybrid and contrapuntal space. Besides the ontological synergies between ethnographic and digital data, the DIGINAUTS research methodologically constructed this space by selecting publicly available online user-generated content, which also acted as an inspirational point of departure for the launch of the ethnographic work (Mollerup and Sandberg, forthcoming). While we acknowledge the constructed nature of this space, we have also shown how it can function as a curated map, producing a partial but relevant representation of the digital actors, which resonates with the offline events. Previous impressive research on migration and digital media has addressed the multiplicity that characterises the use of ICT by migrants before their journeys, en route, and while settling down. This work has diversely described and defined different spaces populated by migratory subjects. This chapter suggests that spatial diversity is also enacted through the methodological multiplicity used to represent space. In onto-epistemological terms, the research methodology for selecting data as well as the data themselves enact the conceptual representation of space.

In our research and methodological context, migrant digital space opens the possibility of a combination of a macroscopic level of analysis and a more small-scale in-depth overview of the (digital) everyday encounters of migrants with digital media. By focussing on longitudinal temporal evolutions (e.g., Fig. 2.2) we observe macro trends and the impact of world-events on digital data, while, at the same time, we can dive into a user-centric perspective by analysing single messages or single pages. Following a relational approach to space, MDS constitutes an online–offline polyphony that implies flows of information, knowledge, advocacy, solidarity, politics, as well as hazards between and for migrants. This multi-layered space consists of digital subjects, migration-related topics, and several different digital objects. These multiple layers simultaneously define and enact MDS. At the same time, given that we linked digital space with specific geolocations (Greece, Germany, and the Øresund region) and temporalities (specific critical moments in contemporary genealogy of migration to Europe), these dimensions also define MDS. In other words, MDS is sensitive to geographical and temporal properties. Therefore, the practice of mapping the MDS reflects that a space at the intersection of the offline–online worlds is simultaneous and polyphonous. This is not a binary but a multi-levelled space. Mapping is not performed as a "technology of power" but as an unfinished practice, an arena of possibilities that disrupt the sense of coherence and of totality (Massey 2005, 109, 120). With that said, we also acknowledge the complexity of this methodological endeavour. Moving beyond

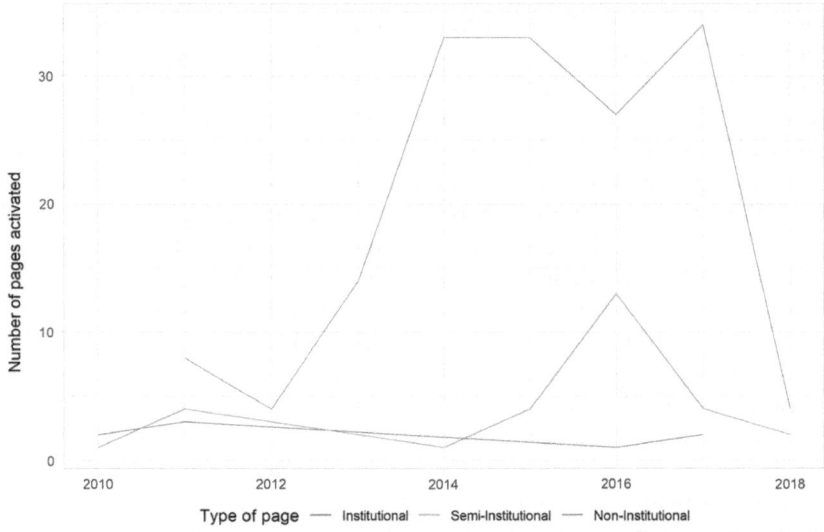

Fig. 2.2 The ephemerality of MDS defines a porous spatiality always under construction where accounts and pages are abandoned and created continuously while new applications enter the arena. This image is used with permission of the authors of this chapter [Rightsholders]

established methodological as well as empirical space brought us to unexplored waters in the sense that we did not limit our research to the official digital voices of formal institutions or actors involved in migration politics. Exploring the MDS also involved taking into account methodological and empirical heterogeneous resources spread across the digital sphere and various digital platforms: commercial social media, websites, and messaging and voice-over-IP services, among others. While not all these sources ended up being "collected data", for the ethical reasons discussed above, they still constitute data that the researchers can, with a range of ethically robust strategies and non-computational methods, integrate within the research. This meant the constitution of the MDS was rather unstable as the digital traces and switches between different media are entirely random and unruly, while the survival of digital subjects was also unpredictable. The MDS evolves in a constant and dynamic state of reshaping.

We therefore suggest four axes for working with MDS: (A) User-centric perspective. This is also compatible with an Autonomy of Migration perspective, meaning that we compose MDS with migrants' digitally expressed interests as a point of departure. (B) Ethical data collection. We show epistemological solidarity with migrants by protecting and respecting their digital privacy as well as potentially controversial information retrieved online that exposes them and their mobility to risk. (C) Temporarily limited. As we saw, accessing data is one thing. Accessing for a longer period is quite another. Even when large online platforms, such as Facebook, make data available, researchers have no guarantee that the data they retrieve is accurate or long-lasting. (D) Alternation of scales. The multifarity of MDS allows us to apply digital methods on several scales, challenging digital commonsensical patterns.

An important element of MDS worth discussing separately here is migrants' everyday digital practices, since they provided us with significant classifications of everyday life practices that were directly projected in digital space. Group chatting, video calling, microblogging, emailing, online games, online music, online shopping, posting, commenting on social media, sharing, (dis)liking, and so on populate an MDS much larger than our data collection would suggest and possibly larger than any future version of the concept. Migrants, in this expanded MDS enact themes, problems, and challenges that they face and confront daily. Some of it trickled down into the public, ethically acceptable, Facebook-centric version of the MDS that we defined, and this enabled us to reconstruct and analyse migrants' everyday life throughout their journeys. Issues related to food distribution, sexuality, job seeking, navigation, legal advice, education, housing, and economic solidarity were themes directly apparent in our material. This also allowed us to interrelate themes with the specific geographical areas under investigation. The content of the everyday lives of migrants, viewed through the lens of the MDS, altered in the different countries under investigation. As we mentioned above, temporality was also an important factor in the constitution of the MDS. Ephemerality of digital subjects, critical events in the period under investigation (such as the long summer of migration and winterisation), differentiated accessibility and availability of data in time depending on digital platforms' policy strategies, and the nature of digital subjects (institutional, semi-institutional, and non-institutional) significantly reshaped the MDS.

Bibliography

Alam, Khorshed, and Sophia Imran. 2015. "The Digital Divide and Social Inclusion among Refugee Migrants: A Case in Regional Australia." *Information Technology & People* 28.

Alencar, Amanda. 2018. "Refugee Integration and Social Media: A Local and Experiential Perspective." *Information, Communication and Society* 21 (11): 1588–1603.

Alencar, Amanda. 2019. "Digital Place-Making Practices and Daily Struggles of Venezuelan (Forced) Migrants in Brazil." In *The Sage Handbook of Media and Migration*, edited by K. Smets, K. Leurs, M. Georgiou, S. Witteborn, and R. Gajjala, 503–513. Thousand Oaks, CA: Sage.

Alevizou, Giota. 2019. "Civic Media and Placemaking: (Re)Claiming Urban and Migrant Rights Across Digital and Physical Spaces." In *The Sage Handbook of Media and Migration*, edited by K. Smets, K. Leurs, M. Georgiou, S. Witteborn, and R. Gajjala, 489–502. Thousand Oaks, CA: Sage.

Alinejad, Donya, and Sandra Ponzanesi. 2020. "Migrancy and Digital Mediations of Emotion." *International Journal of Cultural Studies* 23 (5): 621–638.

Almenara-Niebla, Silvia. 2020. "Making Digital 'Home-Camps': Mediating Emotions among the Sahrawi Refugee." *International Journal of Cultural Studies* 23 (5): 728–744.

Amelina, Anna, Horvath Kenneth, and Meeus Bruno, eds. 2016. *An Anthology of Migration and Social Transformation: European Perspectives*. Cham: Springer.

An Architektur. 2010. "On the Commons: A Public Interview, with Massimo De Angelis and Stavros Stavrides." e-flux (17).

Aouragh, Miriyam. 2011. *Palestine Online: Transnationalism, the Internet and the Construction of Identity*. London: Tauris.

Bork-Hüffer, Tabea. 2016. "Mediated Sense of Place: Effects of Mediation and Mobility on the Place Perception of German Professionals in Singapore." *New Media and Society* 18 (10): 2155–2170.

Borkert, Maren, Karen E. Fisher, and Eiad Yafi. 2018. "The Best, the Worst, and the Hardest to Find: How People, Mobiles, and Social Media Connect Migrants In(to) Europe." *Social Media + Society* 4 (1): 205630511876442.

Bornakke, Tobias, and Brian L. Due. 2018. "Big–Thick Blending: A Method for Mixing Analytical Insights from Big and Thick Data Sources." *Big Data and Society* 5 (1): 2053951718765026.

Brenner, Neil. 2001. "The Limits to Scale? Methodological Reflections on Scalar Structuration." *Progress in Human Geography* 25 (4): 591–614.

Brinkerhoff, Jennifer M. 2009. *Digital Diasporas: Identity and Transnational Engagement*. Cambridge; New York: Cambridge University Press.

Brun, Cathrine. 2001. "Reterritorilizing the Relationship between People and Place in Refugee Studies." *Geografiska Annaler: Series B. Human Geography* 83 (1): 15–25.

Bruns, Axel. 2019. "After the 'APIcalypse': Social Media Platforms and Their Fight Against Critical Scholarly Research." *Information, Communication and Society* 22 (11): 1544–1566.

Certeau, Michel de. 1984. The Practice of Everyday Life. Berkeley, California: University of California Press.

Charmarkeh, Houssein. 2013. "Social Media Usage, Tahriib (Migration), and Settlement among Somali Refugees in France." *Refuge: Canada's Journal on Refugees* 29 (1): 43–52.

Costa, Elizabetta, and Xinyuang Wang. 2019. "Being at Home on Social Media: Online Place-Making among the Kurds in Turkey and Rural Migrants in China." In *The Sage Handbook of Media and Migration*, edited by K. Smets, K. Leurs, M. Georgiou, S. Witteborn, and R. Gajjala, 515–525. Thousand Oaks, CA: Sage.

de Certeau, Michel. 2013. *The Practice of Everyday Life. 1: ... 2*. Print. Berkeley, CA: University of California Press.

Dekker, Rianne, Godfried Engbersen, Jeanine Klaver, and Hanna Vonk. 2018. "Smart Refugees: How Syrian Asylum Migrants Use Social Media Information in Migration Decision-Making." *Social Media + Society* 4 (1): 205630511876443.

Dessewffy, Tibor, and Zsofia Nagy. 2016. "Born in Facebook: The Refugee Crisis and Grassroots Connective Action in Hungary." *International Journal of Communication IJOC*: 2872–2894.

Diminescu, Diana. 2008. "The Connected Migrant: An Epistemological Manifesto." *Social Science Information* 47 (4): 565–579.

Drüeke, Ricarda, Elisabeth Klaus, and Anita Moser. 2019. "Spaces of Identity in the Context of Media Images and Artistic Representations of Refugees and Migration in Austria." *European Journal of Cultural Studies* 1367549419886044.

Farman, Jason. 2021. *Mobile Interface Theory: Embodied Space and Locative Media*.

Georgiou, Myria. 2013. *Media and the City: Cosmopolitanism and Difference. 1.* publ. Cambridge: Polity Press.

Gillespie, Marie et al. 2016. Mapping Refugee Media Journeys: Smartphones and Social Media Networks.

Gillespie, Marie, Souad Osseiran, and Margie Cheesman. 2018. "Syrian Refugees and the Digital Passage to Europe: Smartphone Infrastructures and Affordances." *Social Media + Society* 4 (1): 2056305118764440.

Grzymala-Kazlowska, Alexandra, and Jenny Phillimore. 2018. "Introduction: Rethinking Integration. New Perspectives on Adaptation and Settlement in the Era of Super-Diversity." *Journal of Ethnic and Migration Studies* 44 (2): 179–196. https://doi.org/10.1080/1369183X.2017.1341706.

Gualda, Estrella, and Carolina Rebollo. 2016. "The Refugee Crisis on Twitter: A Diversity of Discourses at a European Crossroads." *Journal of Spatial and Organizational Dynamics* 4 (3): 199–212.

Harvey, David. 2006. *"Spaces of Global Capitalism: A Theory of Uneven Geographical Development."* 1st ed. London; New York, NY: Verso.

Ho, Justin Chun-Ting. 2020. "How Biased Is the Sample? Reverse Engineering the Ranking Algorithm of Facebook's Graph Application Programming Interface." *Big Data and Society* 7 (1): 2053951720905874.

Kaun, Anne, and Julie Uldam. 2018. "Digital Activism: After the Hype." *New Media & Society* 20 (6): 2099–2106.

Kok, Saskia, and Richard Rogers. 2017. "Rethinking Migration in the Digital Age: Transglocalization and the Somali Diaspora." *Global Networks* 17 (1): 23–46.

Laguerre, M. S. 2010. "Digital Diaspora: Definition and Models: Diasporas in the New Media Age: Identity, Politics, and Community." In *Diasporas in the New Media Age: Identity, Politics, and Community*, edited by Andoni Alonso and Pedro Oiarzabal, 49–64. Reno: University of Nevada Press.

Latonero, Mark, and Paula Kift. 2018. "On Digital Passages and Borders: Refugees and the New Infrastructure for Movement and Control." *Social Media + Society* 4 (1): 2056305118764432.

Lefebvre, Henri. 1991. The production of space. Oxford, OX, UK; Cambridge, Mass., USA: Blackwell.

Lefebvre, Henri. 2014. *Critique of Everyday Life: The Three-Volume Text.* Translated edition. London: Verso.

Leurs, Koen. 2014. "Digital Throwntogetherness: Young Londoners Negotiating Urban Politics of Difference and Encounter on Facebook." *Popular Communication* 12 (4): 251–265.

Leurs, Koen. 2015. *Digital Passages: Migrant Youth 2.0: Diaspora, Gender and Youth Cultural Intersections.* Amsterdam: Amsterdam University Press.

Leurs, Koen. 2019. "Transnational Connectivity and the Affective Paradoxes of Digital Care Labour: Exploring How Young Refugees Technologically Mediate Co-presence." *European Journal of Communication* 34 (6): 641–649.

Leurs, Koen, and Kevin Smets. 2018. "Five Questions for Digital Migration Studies: Learning From Digital Connectivity and Forced Migration in(to) Europe." *Social Media + Society* 4 (1): 205630511876442.

Lim, Sun, Tabea Bork-Hüffer, and Brenda S. A. Yeoh. 2016. "Mobility, Migration and New Media: Manoeuvring through Physical, Digital and Liminal Spaces." *New Media and Society* 18 (10): 2147–2154.

Madianou, Mirca, and Daniel Miller. 2012. *Migration and New Media: Transnational Families and Polymedia.* Abingdon, Oxon; New York: Routledge.

Makrygianni, Vasiliki, and Galis, Vasilis. Forthcoming. "Practices of radical digital care: narratives of queer and trans solidarity from Greece."

Margetts, Helen, Peter John, Scott Hale, and Taha Yasseri. 2016. "Tiny Acts of Political Participation." In *Political Turbulence: How Social Media Shape Collective Action*. Princeton: Princeton University Press.

Marino, Sara. 2015. "Making Space, Making Place: Digital Togetherness and the Redefinition of Migrant Identities Online." *Social Media + Society* 1 (2): 2056305115622479.

Massey, Doreen. 1994. *Space, Place, and Gender*. NED-New ed. Minneapolis: University of Minnesota Press.

Massey, Doreen. 1999. "Imagining Globalization: Power Geometries of Time-Space." In *Global Futures: Migration, Environment and Globalization*, edited by A. Brah, M. Hickman, and Máirtin Mac an Ghaill. Cham: Springer.

Massey, Doreen. 2003. "Some_Times_of_Space." In *Olafur Eliasson: The Weather Project*, edited by Susan May, 107–118. London: Tate Publishing.

Massey, Doreen B. 2005. *For Space*. London: Thousand Oaks, CA: Sage.

Merisalo, Maria, and Jussi S. Jauhiainen. 2020. "Digital Divides Among Asylum-Related Migrants: Comparing Internet Use and Smartphone Ownership." *Tijdschrift voor economische en sociale geografie* 111 (5): 689–704.

Miller, Daniel, Jolynna Sinanan, Xinyuan Wang, Elisabetta Costa, Shriram Venkatraman, Juliano Spyer, Razvan Nicolescu, Tom McDonald, and Nell Haynes, eds. 2016. *How the World Changed Social Media*. London: UCL Press.

Mollerup, Nina Grønlykke, and Marie Sandberg. Forthcoming. "'Fast Trusting'—Practices of Trust During Irregularised Journeys to and Through Europe." In *The Migration Mobile: Border Dissidence, Sociotechnical Resistance and the Construction of Irregularised Migrants*, edited by Vasilis Galis, Martin Bak Jørgensen, and Marie Sandberg. Lanham: Rowman & Littlefield.

Moores, Shaun. 2012. *Media, Place and Mobility*. Houndmills, Basingstoke, Hampshire; New York: Palgrave Macmillan.

Papataxiarchis, Evthymios. 2016. "Being 'There': At the Front Line of the 'European Refugee Crisis'—Part 1." *Anthropology Today* 32 (2): 5–9.

Polson, Erika. 2015. "A Gateway to the Global City: Mobile Place-Making Practices by Expats." *New Media and Society* 17 (4): 629–645.

Roseneau, James N. 2003. *Distant Proximities: Dynamics beyond Globalization*. Princeton, NJ: Princeton University Press.

Sánchez-Querubín, Natalia, and Richard Rogers. 2018. "Connected Routes: Migration Studies with Digital Devices and Platforms." *Social Media + Society* 4 (1). Published electronically 1 January.

Shah, Nishant. 2019. "The Cog That Imagines the System: Data Migration and Migrant Bodies in the Face of Aadhaar." In *The Sage Handbook of Media and Migration*, edited by K. Smets, K. Leurs, M. Georgiou, S. Witteborn, and R. Gajjala, 465–476. Thousand Oaks, CA: Sage.

Siapera, Eugenia. 2014. "Diasporas and New Media: Connections, Identities, Politics and Affect." *Crossings: Journal of Migration and Culture* 5 (1): 173–178.

Siapera, Eugenia, Moses Boudourides, Sergios Lenis, and Jane Suiter. 2018. "Refugees and Network Publics on Twitter: Networked Framing, Affect, and Capture." *Social Media + Society 4* (1): 205630511876443

Smets, K. (2018). The way Syrian refugees in Turkey use media: Understanding "connected refugees" through a non-mediacentric and local approach. *Communications* 43 (1): 113–123.

Smets, Kevin, Koen Leurs, Myria Georgiou, Saskia Witteborn, and Radhika Gajjala, eds. 2019. *The Sage Handbook of Media and Migration.* 1st ed. Thousand Oaks, CA: Sage.

Smith, Dorothy E. 1987. *The Everyday World as Problematic: A Feminist Sociology.* Boston: Northeastern University Press.

Stavrides, Stavros. 2010. *Meteōroi Chōroi Tēs Heterotētas.* 1st ed. Athens: Ekdoseis Alexandreia.

Stavrides, Stavros. 2016. *Common Space: The City as Commons.* London: Zed Books.

Trimikliniotis, N., D. Parsanoglou, and V. Tsianos. 2014. *Mobile Commons, Migrant Digitalities and the Right to the City.* Cham: Springer.

UNHCR. 2015. "Emergency Appeal 2015: Winterization Plan for the Refugee Crisis in Europe (November 2015–February 2016)—World." UNHCR, 5 November 2015.

van Dijck, José. 2009. "Users like You? Theorizing Agency in User-Generated Content." *Media, Culture & Society 31* (1): 41–58.

van Dijk, Teun. 2018. "Discourse and Migration." In *Qualitative Research in European Migration Studies, IMISCOE Research Series,* eds. Evren Yalaz and Ricard Zapata-Barrero. Cham: Springer International Publishing: Imprint: Springer, 227–45.

Volkmer, Ingrid, ed. 2012. *The Handbook of Global Media Research.* Chichester, West Sussex: Wiley-Blackwell.

Witteborn, Saskia. 2012. "Forced Migrants, New Media Practices, and the Creation of Locality." In *The Handbook of Global Media Research, Handbooks in Communication and Media,* edited by I. Volkmer, 321–330. Chichester, West Sussex: Wiley-Blackwell.

Witteborn, Saskia. 2014. "Forced Migrants, Emotive Practice and Digital Heterotopia." *Crossings: Journal of Migration and Culture* 5 (1): 73–85.

Witteborn, Saskia. 2015. "Becoming (Im)Perceptible: Forced Migrants and Virtual Practice." *Journal of Refugee Studies* 28 (3): 350–367.

Xie, Zhuoxiao, and Saskia Witteborn. 2019. "The Mobility–Migration Nexus: The Politics of Interface, Labor, and Gender." In *The Sage Handbook of Media and Migration*, edited by K. Smets, K. Leurs, M. Georgiou, S. Witteborn, and R. Gajjala, 453–463. Thousand Oaks, CA: Sage.

Yilmaz, Gaye Gokalp. 2013. "Tactics in Daily Life Practices and Different Forms of Resistance: The Case of Turks in Germany." *Procedia—Social and Behavioral Sciences* 82: 66–73.

Yuval-Davis, Nira. 2007. "Intersectionality, Citizenship and Contemporary Politics of Belonging." *Critical Review of International Social and Political Philosophy* 10 (4): 561–74.

Contrapuntal Connectedness: Analysing Relations Between Social Media Data and Ethnography in Digital Migration Studies

Marie Sandberg, Nina Grønlykke Mollerup, and Luca Rossi

INTRODUCTION

Digital methods and computational analysis have made great progress in recent years in the humanities and social sciences. The integration of digital data with other types of materials into heterogeneous assemblages provides intriguing grounds for further investigation. As discussed

M. Sandberg (✉) · N. G. Mollerup
Centre for Advanced Migration Studies (AMIS), The SAXO Institute,
University of Copenhagen, Copenhagen, Denmark
e-mail: sandberg@hum.ku.dk

N. G. Mollerup
e-mail: ninagm@hum.ku.dk

L. Rossi
IT University of Copenhagen, Copenhagen, Denmark
e-mail: lucr@itu.dk

© The Author(s) 2022   53
M. Sandberg et al. (eds.), *Research Methodologies and Ethical Challenges in Digital Migration Studies*, Approaches to Social Inequality and Difference, https://doi.org/10.1007/978-3-030-81226-3_3

in recent digital ethnographic research (Boellstorff and Maurer 2015; Borkert et al. 2018; Blok et al. 2017; Curran 2013; Gillespie et al. 2018; Munk 2019), ethnographic materials are often treated as "thick data" due to their being generated from in situ research methods such as participant observation and interviews. In contrast, so-called big data, including API-generated social media data, is deemed "thin data" as it is generated on the basis of computational methods, which capture a broader material that does not permit in-depth investigation. In set-ups like this, ethnographic material is likely to function as adding background knowledge or context for other types of data, as discussed by Wang (2013). However, as argued by Ingold (2018, 169), placing other lives—in this case, embodied by our two different kinds of research material—within their social, cultural, and historical contexts is like "laying them to rest, putting them to bed, so that we need no longer engage with them directly. Embedding lives in context implies an already completed conversation." Accordingly, it has been argued that ethnographic materials and "big social data" should be considered not simply as different from but also as complementary with one another and capable of being stitched or assembled into analytical compositions and insights (Blok et al. 2017). While we take these efforts as a point of departure, we also see a larger potential in the work of radically rethinking the relations between different types of materials, a potential that goes beyond either stitching or assembling. In our work with large quantities of API-generated Facebook data (hereafter, social media data) alongside ethnographic materials, we seek to transcend conceptualising the relationship between sets of research material as either *confirming*, *complementing*, or *creating context* for one another. While we acknowledge the existence of other types of big social data (Manovich 2011, 460) (as well as several types of big non-social data), social media data is, as we shall see, among the most commonly used big social data that has been applied to computationally study migrants.

In this contribution, we rethink the relationship between our social media data and ethnographic material by showing them to be fundamentally *interconnected*. We explore the potentials of combining ethnography and social media data by establishing relations in our material as *contrapuntal*. Counterpoint in music occurs when several different musical lines play simultaneously, being at once independent and related.[1] Inspired by Tim Ingold (2018), we understand lines of counterpoint as translated into human movements, which carry on alongside one another, not as the summation of parts but as the correspondence of their particulars.

We understand these related but not necessarily coordinated movements as contrapuntally interconnected. We further unpack some of the challenges inherent in working with social media data and using analytical data processing programmes. These involve, for example, the generation of data through logics of quantification and bias towards numbers of likes and spikes in the material, which Rogers (2018, 450–454) designates as "vanity metrics" (for further critical discussion of social media metrics, see also Tufekci 2014, 505–514; Kitchin 2014, 1–12). In order to avoid such analytical traps, in which data and materials are ordered along predefined scales of big or small, thick or thin, we pursue a strategy of *non-scalability* (Tsing 2012). For Anna Tsing non-scalability implies "letting scale arise from the relationships that inform particular projects, scenes, or events" (ibid., 509). As Strathern has argued, the act of scaling relies on an ontology, in which the world is composed of parts that add up to a whole, suggesting an approach of working upwards from the small details to the big picture (1991, 109f). Translated into the discussions surrounding digital methods, Latour et al. (2012, 591) have demonstrated that, by following the connections between digital traces left in available databases, it becomes possible to transcend this ontology of parts and wholes altogether. Rather than presupposing the two levels of social order—of parts and wholes, micro and macro, elements and aggregates—we follow the approach of working from the middle (Haraway 1988), in which materials are created differently and speak different languages but are nonetheless produced through engagement with the same world. With help from Ingold, we establish contrapuntal rather than summative relations in our materials. People live alongside one another; sometimes they meet, sometimes they move away from each other, yet they correspond with the same world. They are "moving on, alongside one another," which makes them *attentive, responsive,* and *responsible* to one another (Ingold 2018, 160). The task then becomes one of demonstrating the analytical potential of contrapuntal interconnectedness and how different-yet-related data and materials are answerable to the same world. By establishing contrapuntal relations in our material, we explore and qualify this affinity with the aim of identifying further potentials and questions for digital migration research when bringing social media data and ethnographic materials into conversation.

This chapter is based on ethnographic fieldwork carried out in the Danish–Swedish borderlands in 2018–2019 as well as social media data collected through API access from public Facebook pages (including posts and comments) related to irregularised migration and refugee relief in the Danish–Swedish borderlands, including German pages linking to the German–Danish borderlands, covering the period between 2011 and 2018.

In the following, we present our understanding of one world anthropology and contrapuntal analysis. We then position our contribution in the emerging field of digital migration research and discuss how we are in dialogue with and differ from similar attempts at combining big social data with ethnographic materials. We subsequently elaborate upon our methodological approach and the ethical challenges connected to the use of large-scale social media data in the context of migration research. Finally, we test the potentials for conducting contrapuntal analysis. For the purpose of our argument here, which is to propose a contrapuntal analytical strategy for engaging ethnographic material with social media datasets, the analysis will remain illustrative. Because our research is part of the wider DIGINAUTS project (see Sandberg and Rossi in the Introduction to this book), its overall scope focused on how irregularised migrants and solidarity workers challenge the European border regime through their respective digital fields of navigation. *The border* thus emerged as a recurring point of tension in our ethnographic conversations, in the social media dataset, and accordingly in our analytical and theoretical discussions. The contrapuntal analysis presents three different versions of the border enacted through the material, when conversing between the different-yet-related datasets and materials. In conclusion, we discuss how the contrapuntal move can advance digital humanities and the field of digital migration studies, not only in methodological but also in analytical and theoretical ways.

One World Anthropology

In acknowledging the interconnectedness of our ethnographic material and social media data and bringing them into conversation without positioning them as each other's stand-ins or contextual backgrounds, we find inspiration in Tim Ingold's (2018) *one world anthropology*. Ingold challenges the idea of life-as-a-whole as a sum of its parts and proposes the idea of correspondence, which entails that "parts are not components that are added *to* one another but movements that carry on *alongside* one another, so too, in the human family, lives lived in counterpoint are not 'and … and … and' but 'with … with … with.' And in answering—or responding—to one another, they co-respond" (160, emphasis in original). He contends that "life itself, then, is not the summation but the correspondence of its particulars" (158). Ingold argues that this calls for "a 'turn' that is not ontological but ontogenetic" (169), that is, a turn which focuses on the ongoing generation of being rather than its essence (167). And this "leads us to conceive of the one world as neither a universe nor a fractiverse but as a pluriverse" (169). As mentioned, Ingold proposes the analogy of music, where the "relation between parts and whole is not summative – neither additive nor multiplicative – but *contrapuntal*" (160, emphasis in original). Ingold thus thinks of "the life of every particular soul (...) as a line of counterpoint that, even as it issues forth, is continually attentive and responsive to each and every other" (160).

The idea of contrapuntality was introduced to cultural studies by Edward Said (1994, 2000). It is especially pertinent for us to (re)turn to Said since the state of exile, which is at the heart of our endeavour, was crucial to his life and scholarship. Said's idea of contrapuntality is based on the same premise of connectedness as that of Ingold. Relations between coloniser and colonised, former coloniser and formerly colonised take centre stage in Said's reflections. Said contends, "we must be able to think through and interpret together experiences that are discrepant, each with its particular agenda and pace of development, its own internal formations, its internal coherence and system of external relationships, all of them co-existing and interacting with others" (1994, 32). With the musical metaphor, "various themes play off one another, with only a provisional privilege being given to any particular one; yet in the resulting polyphony there is concert and order, an organized interplay that derives from the themes, not from a rigorous melodic or formal principle outside the work" (Said 1994, 51). As Said suggests, we must not foreground any composer or mastermind behind "the music"; it is necessary to remind ourselves that there is always a limit to the metaphor. For Said, contrapuntality becomes a method for simultaneous awareness of different voices, one that acknowledges their connectedness and answerability to each other and that allows alternative or new narratives to emerge (1994, 51).

Life as Experienced, Traced Life

While Ingold, working from the perspective of anthropology, speaks of lives and souls, Said, grounded in the tradition of literary criticism, speaks of texts and voices. We do not regard these contrapuntal understandings as contradictory but instead find it useful that the two foci explicitly address our two types of material, namely the lived lives that take centre stage in our ethnography and the texts that we choose to foreground in our social media data. That is, while we maintain that our materials are inherently interconnected—corresponding with the same world—we remain aware that the different modes of data production have facilitated different paths to knowledge: one foregrounds life as experienced, and the other foregrounds textual traces.

Before elaborating upon why we find contrapuntal analysis particularly relevant for our analysis of disparate data and material, it is useful to expand upon how we perceive these data and materials. We cautiously designate the content we collected through Facebook APIs as *data*. This is because the data is actually produced by the API on the basis of activities that take place on the platform (Lomborg and Bechman 2014, 256) and because this term is commonly used in computational sciences. When we refer to our ethnographic observations, reflections, and transcripts as *materials*, we do so in recognition of Ingold's (2013, 5) understanding of ethnography as a way of "knowing from the inside." Ingold contends that:

> to convert what we owe to the world into 'data' that we have extracted from it is to expunge knowing from being. It is to stipulate that knowledge is to be reconstructed on the outside, as an edifice built up 'after the fact', rather than as inhering in skills of perception and capacities of judgement that develop in the course of direct, practical and sensuous engagements with our surroundings. (Ingold, 2013, 5)

Approaching our social media data with an ontological commitment to *knowing from the inside* entails recognition of these as actively produced in correspondence between our research team's decisions, social media platforms and programme logics, irregularised migrants, solidarians and other Facebook users, writings and movements, and more. That is, despite their diverse ontological heritages and the frictions these differences produce, our social media data and ethnographic materials have a fundamental affinity that we wish to bring into focus through a contrapuntal analysis.

Our point is not to seek out particular connections between our data and material but to instead acknowledge their *interconnectedness*. The texts of our social media data are fragmented and partial, but they are produced relationally with the ethnographic settings we have explored, in Ingold's words, "*with...with...with.*" As we bring different elements together in a contrapuntal analysis, we allow new narratives—non-singular and situated—to emerge. The contrapuntal approach highlights a tension between on the one hand recognising materials and data as fundamentally interconnected and on the other hand recognising that the researcher actively composes connections. Pursuing a contrapuntal analysis, we consciously position ourselves within this tension (cf. Haraway 1988).

Contrapuntal analysis is particularly relevant to our material because it not only acknowledges the interconnectedness of lives but also speaks to the particular historical moment of 2015, when the stream of irregularised migrants to and through Europe brought people together and strengthened connections between people who had never before had direct engagements. Rather than being detached fragments, disconnected from people's experiences, our social media data speaks directly to how many of our research participants experienced this time, a time when precisely these types of digital messages frequently proved crucial to the course of their lives. It was a moment that few of the people who were directly involved experienced coherently or cogently. This shared moment illustrates the affinity between our material and data, and it is exactly this co-existence between incoherence and instability—and connectedness—that we wish to capture with the contrapuntal approach.

On Non-Scalability

In her article "On non-scalability," Tsing (2012, 507) defines scalability as the ability to expand without rethinking the basic elements, features, or designs of a project: "To 'scale up', indeed is to rely on scalability – to change the scale without changing the framework or knowledge of action." Tsing also presents the notion of *precision nesting* of scales, in which "the small is encompassed neatly by the large" without any revisions to the design or nature of the project (ibid.). Whereas scalability is a tenet of capitalism's growth ideology, which promotes the idea of endless growth of businesses along the same scale, *non-scalability* presents the idea of "letting scale arise from the relationships that inform particular projects, scenes, or events" (ibid., 509). "In that work, there are big stories as well as small ones to tell. There is no requirement that the scales nest or that one performs wizardry of conversion from one to the other without distortion" (ibid., 509f). While tracing contrapuntal connectedness, the question arose how to build an understanding of the phenomenon that "scales smoothly from minute details to aggregate patterns and back" (Munk 2019, 169). In other words, through our contrapuntal move, we find it crucial to go beyond the particular case, albeit in non-scalability mode.

Digital Migration Studies: An Emerging Field

Digital migration studies has emerged as a relevant field for providing new and meaningful insight into the phenomenon of human migration and migrants' practices (Kok and Rogers 2017, 23–46; Leurs and Smets 2018). Following the rapid adoption of digital material and methods in the humanities and social sciences, digital migration studies has developed in numerous, often only loosely connected directions. While there is fundamental agreement that digital media and digital technologies repurpose—and frequently facilitate—the process of migration (Diminescu 2008, 565–579) and affect migrant populations' processes of social integration and political participation (Komito 2011, 1075–1086), there is considerable diversity in research foci, empirical foundations, and methods (Leurs and Smets 2018).

Calls have been made recently to systematise existing approaches by further reflecting upon methodological implications of studying migrants' digital practices. Leurs and Prabhakar (2018, 247–266) map the field of digital migration studies by identifying three distinct paradigms: (I) migrants in cyberspace, (II) everyday digital migrant life, (III) migrants as data. These three existing paradigms include main representative scholars, alternative theoretical discourses, and often non-overlapping methodological preferences. While the first paradigm is rooted in the hermeneutical approach to digital humanities, the second builds on social science theories and methods that precede digital data (e.g., ethnography, interviews, and participatory observation). Finally, the third paradigm is organised around the idea of digital methods as research methods and practices uniquely tailored to handling contemporary digital traces.

Following the call to establish bridges between these different paradigms, we situate our research in conversation with the non-digital-media-centric ethnographic approach, which focuses on everyday digital migrant life (paradigm II), and the digital-media-centric digital approach, which appreciates *migrants as data* (paradigm III). We do so by developing a research design that is from the start equally rooted in social media data and ethnography. In contrast to common practice within mixed methods perspectives (Creswell 2014, 1), we do not define a convergent or fully sequential research design (Snelson 2016) in which one type of data complements or drives the analysis undertaken upon the other type of data. As will be detailed in the methods section, ethnography, data production, and analysis were carried out in relative independence following an initial phase of alignment. For instance, previous knowledge and preliminary field studies informed the initial digital seed for the data production, but data production subsequently followed its own internal process for selecting relevant sources.

Stitching, Complementing, Remixing...

Moving beyond the migration context presented here, several existing studies have approached social media data alongside qualitative or ethnographic materials (Gillespie et al. 2018; Borkert et al. 2018; Curran 2013). Boellstorff and Maurer (2015) bring together anthropological and comparative insights with "big data," highlighting the complexity of enabling conversation between highly divergent methodological approaches and epistemological perspectives. Among recent attempts to combine different data formats and materials, the focus has been on "assembling" or "stitching" heterogeneous data worlds (Blok et al. 2017; Blok and Pedersen 2014). Blok and Pedersen outline a research collaboration between anthropologists and sociologists experimenting with combining big transactional data (based on GPS and Smartphone Bluetooth signals) and ethnographic fieldwork (fieldnotes) into a joint social network analysis. The core metaphor used to build together the data formats is stitching "as a possibility of mutual fertilization across apparently incommensurable fields" (ibid., 2), which recognises the relationship between the two as complementary, flagging an ontological ground of *assemblage* (cf. Carter 2018). The aim is to create an experiment in "'big data-ethnography' in which the hyphen suggests an initial, inherent, and deliberate uncertainty as to which of the two terms (big data, ethnography) is here the context for the other; which is figure and which is ground" (ibid., 4). A further example is provided by (Markham et al. 2013), who, though not focusing specifically on social big data and ethnography, suggests a method of remixing data materials as "a powerful tool for thinking about qualitative, interpretive research practice" in order to "better grapple with the complexity of social contexts characterized by ubiquitous internet, always-connected mobile devices, dense global communication networks, fragments of information flow, and temporal and ad hoc community formations" (65). Markham focuses on the complex manner in which different types of material can be brought together to create particular frames of interpretation. However, the remix metaphor also encompasses a practice of cutting, copying, and pasting (cf. Markham 2017). The contrapuntal approach differs in that it does not embark on "cutting" elements and pasting them into new contexts for new purposes, seeks instead to shift the gaze in order to allow new meanings to emerge from already-connected material. It also differs in highlighting the material types as already connected, rather than focusing on the researcher's remixing or stitching practices. That is, in a contrapuntal analysis, the researcher establishes ways of working analytically with connectedness, as detailed below.

In the specific context of conversing with big data and ethnographic materials, we contend that we are not simply dealing with data of different scales in which big data indicates a sum of the ethnographic material but a more meticulously described part. Anders Munk (2019) argues that, across the "great divide" (cf. Latour 2005) between qualitative and quantitative analysis, a shared disconnect exists among ethnographers and digital researchers. This is because both fields struggle to chart the "native's point of view" into more quantifiable formats while maintaining the intended meaning and original context of those "on-life traces" made in the digital world: "These questions are not simply about making sense of data but about making sense in a way that claims to reflect a native point of view" (Munk 2019, 163). Apart from the complementary model pursued by (Blok and Pedersen 2014), Munk suggests three additional modes for further overcoming the quali-quantitative divide: single-level analysis, curation, and algorithmic sense-making: Whereas single-level analysis sees onlife traces as embodying both qualitative richness and quantifiability, curation presents itself as a critical practice scrutinizing different media environments. Algorithmic sensemaking tends to emulate qualitative sense-making to be replaced by quantitative pattern recognition. (Munk 2019, 164f).

While we understand digital and ethnographic material to originate from the same world, we move beyond the complementarity model. We thus position our approach on a possible overlap between single-level analysis and curation, as proposed by Munk (ibid.). We manually curated a digital data collection of pages—bearing in mind that this was ultimately mediated through Facebook's API and involved repurposing the API for our research goals—to avoid fully automated data crawling and to build a corpus with a purpose (cf. Munk 2019, 172). We simultaneously approached the observed digital data and the quantitative trends not as something to be complemented with our sense-making obtained from ethnographic material but as a means of acknowledging a connectedness that transcends the particular case. We can perhaps see the contrapuntal approach as a fifth mode for overcoming the quali-quantitative divide, a mode in which digital data and ethnographic material can produce insights that are not necessarily complementary but nevertheless refer to the same world.

Our understanding of contrapuntal analysis specifically goes beyond an idea of assembling. We approach our materials neither as (interchangeable) complementary add-ons to one another, as confirmations of one another, nor as contextualisations of one another, regarding them instead as *answerable* to one another, as fundamentally interconnected. We contend that the complementarity approach, alongside the confirmatory approach and contextualisation, fundamentally assume the ability of one type of data to "fit together" with alternative types of data within a larger representation of the world, either by providing an alternative perspective, supporting evidence, or background information. This assumes known or expected relations between the data (how the data types speak to each other) that cannot always be taken for granted.

Methodological Design and Ethical Considerations

The general methodological design of this research builds upon two parallel data sources: social media data generated through API access to a curated list of public Facebook pages and ethnographic material produced through participant observation and interviews. The production of digital data was initially informed by early pilot studies and pre-existing knowledge, while the ethnographic study was at times inspired by insights from our digital data production. However, the actual definition of the digital data sources was not made to mirror or complement the ethnographic observations (see also Makrygianni et al. Chapter 2 in this book). Our ethnographic material was produced through ethnographic fieldwork carried out with Syrian refugees and solidarians in and around the Danish–Swedish borderlands, the Øresund Region, in 2018 and 2019. The research participants, who were refugees, had arrived in Denmark and Sweden as irregularised migrants between 2014 and 2016 and had subsequently obtained juridical refugee status. We therefore maintain the designation *irregularised migrants* since our conversations with research participants concerned the time when they fled, travelling to the Øresund Region without being able to depend upon regular modes of travel. Taking our cue from Rozakou (2018), we use the term *solidarians* to highlight that the various refugee support initiatives included in this study were not necessarily part of established NGOs; they were instead constituted by informal networks of people in the vicinity of the locations where the irregularised migrants arrived, with a common aim of "standing up in solidarity."[2] Key topics in our study were how these irregularised migrants had navigated on their journeys to and within Europe, the role of solidarians in this navigation, and the significance of digital practices for both groups. We therefore used mainly in-depth, retrospective interviews. In our interviews with refugees, we looked back on their journeys, and in our interviews with solidarians, we revisited their work to help irregularised migrants. We interviewed several members of the same family networks in order to hear different accounts of shared journeys and establish trust (Mollerup and Sandberg, forthcoming). This also provided us with detailed insight into the significance of family members who had fled previously.

We interviewed some people on multiple occasions, and in some conversations, we included what we term *device tours* (Mollerup 2020), in which interviewees showed us old conversations, pictures, etc. on phones and computers, allowing these to become focal points in the interviews. These device tours at times played a role in bringing up memories and emotions. For ethical reasons, the contents of our research participants' devices were never systematically recorded, archived, or documented beyond our fieldnotes. Some interviews lasted half an hour, while others lasted several hours and were interwoven with meetings and socialising with family members and others. We carried out interviews in Danish,

English, and Arabic. In total, we undertook 16 interviews with 12 refugees, and 16 interviews with 16 solidarians situated on both sides of the Danish–Swedish border. After transcribing all interviews,[3] we manually coded the transcripts and field notes (526 pages in total) for different themes that had emerged through the fieldwork and through reading and re-reading material. We ended up with 97 codes. One of these codes was "border." The notes and transcripts that were coded "border" deal with stories of irregularised migrants planning to cross, crossing, or failing to cross borders and of solidarians helping irregularised migrants. Many of these stories describe extreme danger and difficulty. They also at times describe remarkable creativity and unexpected success in being able to move.

The digital data production followed a relatively well-established procedure in digital methods: data selection, data enrichment, and data analysis (Kok and Rogers 2017; Rogers 2013). The starting point was constituted by a set of public Facebook pages focused on refugees and irregularised migrants, identified through a combination of methods, including manual snowballing from relevant pages previously known to the research team as well as through Facebook's search function, using topic-specific keywords in English, Arabic, German, Greek, Swedish, and Danish. This produced 200 Facebook pages, which were then manually coded with additional information, such as the type of actor behind the Facebook page, physical location of the actor, date of page creation, and language (or languages) used. All the publicly available content (posts, comments, and reactions) on the pages was downloaded using Facebook's API. This produced a final dataset comprising 200 pages, 84,359 posts and 2,254,923 comments, produced between 20/12/2010 and 24/09/2018. A detailed description of these elements and of the methodological consequences can be read in Chapter 2 (Makrygianni et al.) of this volume. Given the specific focus on the Danish–Swedish border region, we filtered the data so that it contained only pages created by actors from Sweden or Denmark or German pages that connected with the German–Danish borderland. This narrower dataset comprises content posted on 80 public Facebook pages (49,162 posts and 1,238,794 comments). The language distribution is illustrated in Fig. 3.1.

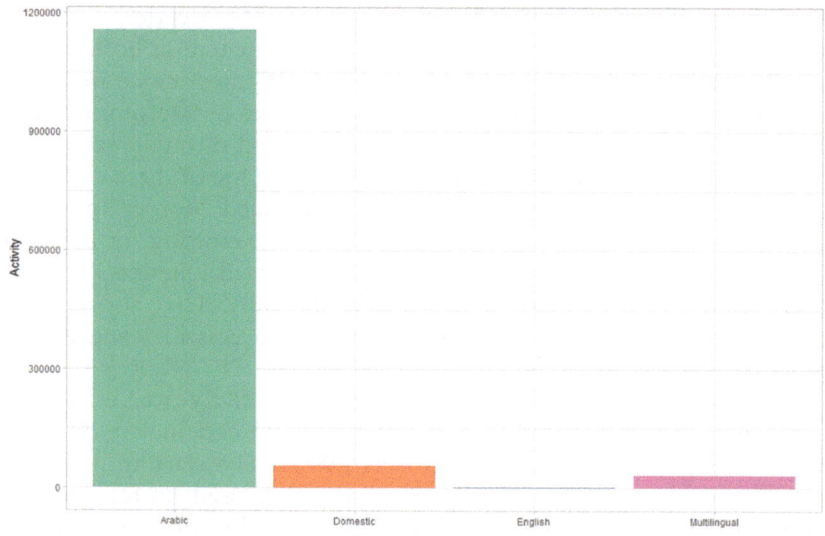

Fig. 3.1 Distribution of languages in social media data, based on number of comments. "Domestic" includes Swedish, Danish, and German (*Source* This image is used with permission of the authors of this chapter [Rightsholders])

On the Ethics of Digital Data for Migration Studies

Use of large-scale social media data in the context of migration research raises important ethical questions. When using digital traces produced by irregularised migrants, we are dealing with a type of material that has been collected without explicit authorisation from the subjects (who at times are unaware of the possibility of data collection). At the same time, gathering and storing large amounts of such data creates additional risks when the practices documented in the data—e.g., undocumented border crossing—are illegal in several countries. While a full analysis of these problems is beyond the scope of this chapter and touches upon several methodological (Lomborg and Bechmann 2014), legal (Kotsios et al. 2019), and policy issues (Bruns et al. 2018), it is possible to adopt specific research practices to mitigate such risks. In the context of the present research, digital data has been collected only from publicly accessible Facebook pages (private and public groups and user profiles were deliberately excluded). In this sense, it is important to stress that our selected social media data from publicly available groups represent only the tip of the iceberg, given that non-public information (including private messages on Messenger, WhatsApp, and similar platforms as well as posts and comments in non-public groups) is not included in the research material for ethical reasons. Identifiable information about the authors of the messages has not been collected at any stage and is not available to the researchers. Furthermore, the data is securely stored and is not accessible to anyone outside the research team. While this somewhat conservative approach is important to secure the anonymity of and prevent harm to people who are active on social media and who never agreed to participate in our research, there is also an important ethical argument for researching this particular historical moment in order to document, analyse, theorise, and contribute to discussions about the EU border regime and its consequences. Our research design, combining computational methods with ethnography, ensures that we can attain a deeper understanding of this moment, acknowledging both the ethical need to research this moment and the ethical need to do so without compromising privacy.

Creating a Data Environment Relevant for Contrapuntal Analysis

In order to be able to work contrapuntally with the two different-yet-interconnected types of research material, we first needed to establish a sustainable data environment relevant for our purpose (cf. the Introduction to this book). Our efforts to approach these two sets of material contrapuntally was challenged by the inbuilt logic in the social media material, which prioritised an approach focused on comments and shares, which encouraged us to search for spikes in activity (Rogers 2013). Looking at intensities in activity enabled us to identify certain spikes that probably related to events we knew to be ethnographically significant (such as the spread of the images of Alan Kurdi, which strongly influenced one of our interviewees' decision to pack his bags and travel to Lesvos, and the advertisements published in Lebanese newspapers by the Danish Minister of Integration at the time, Inger Støjberg, which made one irregularised migrant with whom we spoke reconsider Denmark as a destination). However, these spikes told us little about how these events mattered to the people in question. In addition, we could not assume a relationship between spikes and significant events. For instance, one large spike we looked into in the Swedish material for October 2015 was produced by comments mainly on two particular posts, in the same Arabic language group (we saw a similar spike in the sharing of these two posts). One was a rather humorous video of a Lebanese man in Germany trying to learn Syrian dialect. The other was a lively video of Syrians teaching Germans how to dance the Syrian folk dance *dabke* (see Fig. 3.2). These types of videos were posted on pages that also carried very serious messages and images detailing the devastation of war, fleeing conflict, and quests for information about how to reach safety. Spikes at times corresponded with and contained posts about major events, but the occurrence of a spike did not necessarily provide information about irregularised border crossings.

Seeking instead to move beyond a quantified logic of relevance, in which a relevant period of time would be identified through a peak in the data, and to create a sustainable data environment relevant to our purpose, we filtered the data so that it contained only the period from

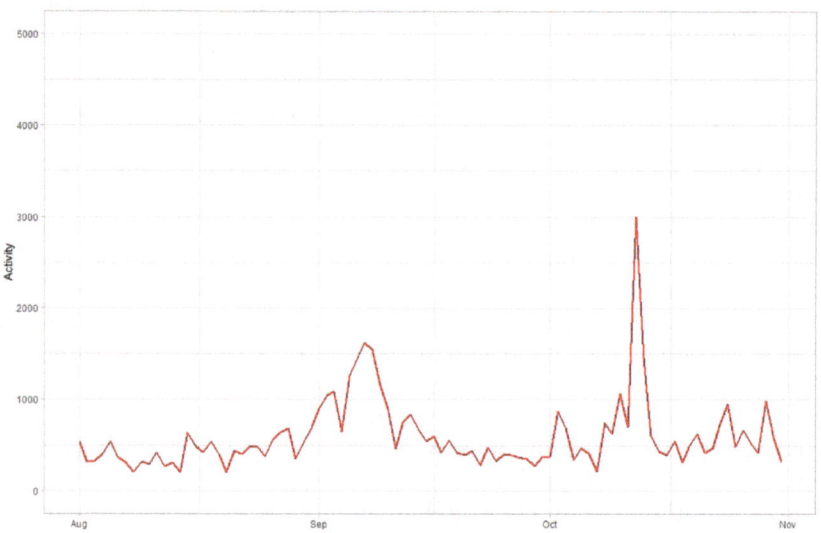

Fig. 3.2 The lure of spikes: Activity in comments in the Swedish material. While at times relationships between events stood out in our ethnography and spikes in activity, we often found attention to spikes to be counterproductive (*Source* This image is used with permission of the authors of this chapter [Rightsholders])

August 2015 to October 2015 and contained only posts. This led to a smaller dataset containing 57 pages and 3456 posts. The period was chosen because it coincided with the arrival of unprecedented numbers of irregularised migrants in Denmark, Sweden, and Europe as a whole, including many of the irregularised migrants with whom we spoke. Correspondingly, a vast array of solidarity initiatives emerged during this period. Many of the solidarians with whom we spoke became active during or immediately in advance of this period. We focused only on posts because comments often strayed far from the topic of the posts, and it proved unfeasible to reconstruct the conversational dynamic between often apparently disconnected comments. Moreover, a focus on posts alone made the quantity of material more manageable. A negative consequence of this decision is, however, a reduction in the diversity of voices, prioritising those individuals who were actually able to and wished to post on the pages.

Having delineated a subset of our digital data, we approached it with the keyword "border." As mentioned above, this recurring keyword was chosen because of our overall research focus on irregularised migrants' and solidarians' strategies for navigating the European border regime through their respective crossings of borders and refugee aid initiatives. Using the Tableau data analysis software, we searched for any occurrence of the word *border* in Danish, Swedish, German, Arabic, and English in our subset of data. Overall, the data contained a reasonably small number of posts containing explicit reference to *border* in the various languages (10 in Danish, 23 in Swedish, 13 in German, 86 in Arabic, 15 in English). Using our ethnographic material as a backdrop, we read and manually coded these posts, paying attention to the nuances in the various references to border, with the aim of conceptualising them as different *versions* of the border (Mol 1999; Law and Urry 2005; Sandberg 2016). We then approached our ethnographic material, which we had manually coded for occurrence of the border concept (not, as with the social media data, for occurrence of the word itself). We extracted those parts of the ethnographic material that were coded with border and then manually coded them for different versions of borders, as we had with the social media data.

We thereby created a sustainable data environment relevant for carrying out contrapuntal analysis, allowing us to simultaneously acknowledge the interconnectedness between our data and material while also allowing us to juxtapose these, enabling productive tension between the ways in which our materials were already connected and the connections we established. Our goal here was not to establish direct relations between our different materials. Instead, we continually approached the two types of materials, seeking to understand how they corresponded with one another (see Fig. 3.3).

During the process, three different but related versions of the border emerged as significant for this chapter. We designate these: *politico-legal border*, *solidary border*, and *border navigated*. Unsurprisingly, these three

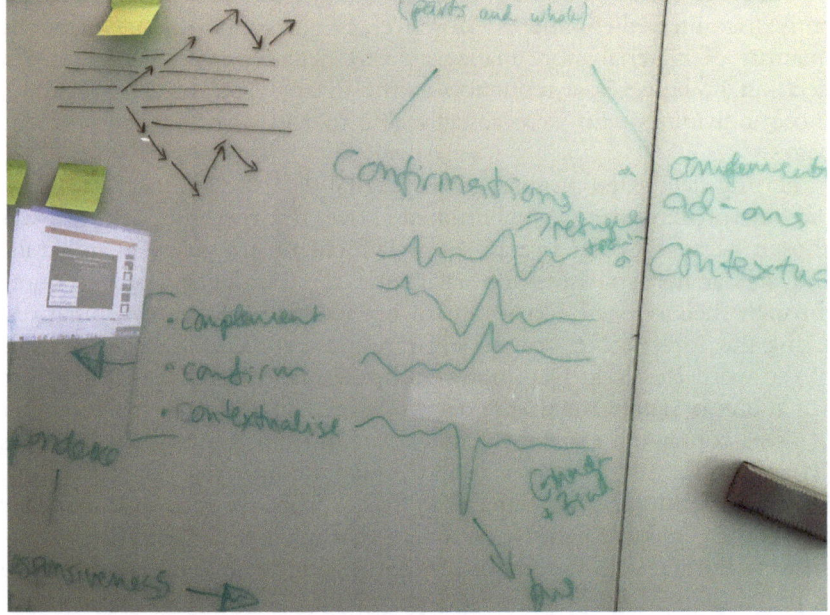

Fig. 3.3 Contrapuntal correspondence. Rather than identifying direct relations between our different types of materials, we aimed to establish correspondence. In the analytical process, we tried to visually illustrate this relationship of "moving alongside one another" through arrows and waves, which react with each other without clashing, meeting, or otherwise making direct contact, and also without necessarily being unidirectional in their movements. The peaks in the top three waves and the corresponding nadir in the bottom wave illustrate when irregularised migrants were told to get on a designated train in Germany, which many did, but which prompted suspicion in one couple with whom we spoke and caused them to choose a different path (*Source* Photo by Nina Grønlykke Mollerup)

versions of the border did not emerge equally from the two sets of material and from the different subgroups of our material. For instance, our Arabic-language social media data, irrespective of country, particularly spoke to *border navigated* and *politico-legal border*, while English and local languages in the respective countries particularly spoke to *solidary border* and *politico-legal border*. Our ethnographic material from conversations with solidarians spoke to all three border versions, while our ethnographic material from conversations with refugees rarely spoke to the solidarian border. While each subgroup of our research material tells stories about the border independently from the others, our analysis emerges from both the social media and the ethnographic material. The ways in which the different border versions formed through and within the different subgroups is thus crucial to our ability to engage in contrapuntal analysis.

Border Enactments

The following analysis serves to illustrate how a contrapuntal lens can help in understanding social media data and ethnographic material as different-yet-interrelated materials. This lens enables us to identify the three different versions of the border enacted through social media communication as well as in situ practices. We present them here first through some of their general characteristics, followed by an exemplary analysis, which simultaneously deploys a contrapuntal approach to our research material and uses the border versions as an analytical lens.

Politico-legal border is concerned with the regulation, politicisation, and control of borders. It includes contestations of both physical and legal borders, recognises borders as changeable, and includes visions for new legal models for a borderless world and longer term plans for managing and controlling borders. Across our research materials, the border emerges as a particular site of struggle (cf. Hess and Kasparek 2017) through calls to action at particular borders and through questioning the moral implications of borders. However, the politico-legal border focuses on borders as a matter of political negotiation and is also addressed in places well removed from physical border crossings, including parliaments and other sites of political debate. *Solidary border* focuses on border-spanning activities that aim to ease irregularised migrants' crossing of borders through refugee relief, everyday humanitarian aid, and transit assistance. Solidary border also includes negotiations of particular borders, for instance, through demonstrations and advocacy for opening borders and organising border crossings. Solidary border works in opposition to several "significant others" such as against racists and related neo-Nazi movements. **Border navigated** deals with practices of illegalised border crossing and thus takes into account the irregularised migrants' perspectives. It attends to different types of border crossings and includes attention to political, social, and material circumstances (weather conditions, landscape, border fences, etc.). Unlike *politico-legal border*, it does not deal with longer term perspectives or moral implications but is instead invested in the here and now of the border and how it might facilitate or hinder movement. Together, the three versions of the border advance an argument concerning *the border multiple* (Andersen and Sandberg 2012), which highlights borders as practice and thus as constituted through a multiplicity of actors while reaching beyond the individual action, site, or event. When we say that borders are *enacted*, we do so to refer to the different ways in which borders are negotiated in experienced life (as depicted in ethnographic conversations) and in traced life (as documented in social media data).

Example #1: The Border as an Event

Our social media data and ethnographic fieldwork focused on the Danish–Swedish borderlands, but it was clear through both sets of material that the border extended well beyond this and was entangled with other borders and solidarity practices spanning the wider European border regime. Thus, the Serbian–Hungarian border emerged as significant in both our social media data and our ethnographic material in the time leading up to Hungary completing the construction of a border fence on 16 October 2015. In the context of this emergence, the different sections of our two sets of material come to have particular significance, as the different data and interviews in different languages with former irregularised migrants and solidarians presenting us with different paths towards understanding the events. A contrapuntal approach allows us to show how the two sets of material present different-yet-related paths to knowledge about irregularised border practices and solidarity actions at this border and beyond, while ensuring a non-scalable mode of working. We see scale as arising from the relationships informing the events or stories that have been told or the words that have been posted or said.

The following information posted in an Arabic-language Facebook group speaks to *the politico-legal border* but can simultaneously affect navigational decisions about where and how to cross borders.

[Social media data] As of tomorrow, Tuesday, Hungary will tighten border crossing procedures and punish illegal crossings with imprisonment. They have sent more than 900 additional police officers to protect the border and close it completely. Posted on Arabs in Denmark, 14 September 2015, translated from Arabic.[4]

A related call for human presence at the border, posted in an English-language German group, shows the *solidary border* as a particular site of negotiation that spans well beyond the actual site:

[Social media data] Please inform everybody who can come should come NOW to the serbian - hungarian border (Horgoš border crossing)! Everything, everything is needed but human presence the most! Thousands of people are demanding without a break the opening of the border! Stand up for solidarity against borders and repression! Posted on Refugees Welcome (Germany), 1 October 2015.

Such calls, public as well as private, have contributed to prompting solidarians with whom we have spoken to radically change their lives by dedicating themselves entirely to helping irregularised migrants for many months. This emerges through our interviews with solidarians. Some could name the precise post, at times including an image, which caused them to get involved. Jens was one of the solidarians who placed his life on hold for months to respond to the situation, including at the Serbian–Hungarian border. He told us:

[Ethnographic material] So, after that, then I was in Presevo [Serbia], driving some emergency aid down there and trying to help out. I had some Danish contacts, who had been down there, when it was complete chaos with 10,000 people [arriving] a day, and there was just no infrastructure or anything. People were soaked in the rain and with two degrees cold [Celsius]. So, I went down there with the car packed with tools and - we just took blankets and children's clothes and everything we could get our hands on and threw them into the back of the car and drove down there. And smuggled it in because we weren't allowed by the authorities in Serbia. Jens,[5] Danish solidarian, 21 March 2019, translated from Danish.

The situation at the Serbian–Hungarian border emerges as chaotic and overwhelming through our social media data and interviews with solidarity workers. Thousands of people are at the border, in bad weather conditions, and with little organisation. Shifting the focus to the *border navigated* provides us with a different view of the situation. When approaching the border through our ethnographic interviews with irregularised migrants, the chaos is still visible, but it is backgrounded, whereas the danger and uncertainty of the situation are highlighted, along with the actual decisions that were made. That is, the chaos is a temporary obstacle that recedes as soon as it is overcome. Most irregularised migrants with whom we spoke singled out Hungary as the most dangerous place or one of the most dangerous places they had encountered in Europe. One irregularised migrant with whom we spoke had been imprisoned and had his belongings "confiscated," and others had, themselves, heard similar stories. Ghada and Ziad, a young couple travelling together, in separate interviews told us of the situation at the Serbian–Hungarian border, which they crossed in mid-late September 2015. Ziad said:

> [Ethnographic material] The weather was so bad we couldn't continue, like, raining so much. So, then we took a bus again, to a city between Hungary and Serbia. I forgot the name. Then everything was hard, because on the way, I mean, the army, they were on the way, and they took people and you have to sign so that means you cannot go to Sweden if you sign in Hungary. Ziad, Syrian refugee, 3 March 2019.

Ghada also told us of their difficulties at the Serbian–Hungarian border and the challenges with the weather:

> [Ethnographic material] I remember, Hungary borders, it was really difficult. This I won't forget. It was really bad. And even the police there were really mean and they almost hit people, like this, 'Just go back! Go back!' They weren't nice. And then we had to walk at night, and we had to wait two nights, I think, two nights. The first night it was raining so [the smugglers] said, 'No, we can't go like this, it's a bit difficult, and we can't find the way.' Ghada, Syrian refugee, 20 March 2019.

Ghada explained how they had finally managed to cross with the help of a smuggler, escaping a volatile situation at the border, where people were being pushed, and they had to hide from the police to avoid getting caught. She elaborated:

[Ethnographic material] We knew that something was happening there in front of us. We didn't know what it was exactly. And we didn't want to just go there. So, we just followed the people who were going on the side. In the bushes. And then we contacted some guy, I think Ziad contacted him. (...) It was a lot of people and I was, like, afraid, I would say. I was really afraid. Ghada, Syrian refugee, March 20, 2019

This ethnographic material highlights the danger and uncertainty experienced at the border; *the border navigated*. For solidarians, this uncertainty was, of course, experienced very differently and instead played into their planning of relief work, allowing what were often much-needed rests and breaks. As a Swedish solidarian, working in a solidarity space that housed irregularised migrants upon immediate arrival, said:

[Ethnographic material] A border would be closed here and there. In Germany, and somewhere else, Serbia for example. We could be told: 'Now the border is closed; it is expected to be opened again in two days. Make sure to rest and sleep now, fewer will come, and later everyone will come again.' And later, we could then be told: 'Now they have reopened. The men have gone first. In the second wave, women and children will come. If you need to collect prams and such, you will need to do it within three days' Like that. So, it was extremely effective. Åsa, Swedish solidarian, 21 March 2019, translated from Swedish.

Approaching the events surrounding the Serbian–Hungarian border around September 2015 contrapuntally through our different materials and their subsets allows us to show how chaos, danger, and uncertainty are differently foregrounded and configured. Our point here is not to seek out direct relations but instead to show contrapuntal correspondence between the different aspects of irregularised border crossings, spanning well beyond individual people and places.

EXAMPLE #2: THE INSTABILITY OF THE BORDER

Irregularised border crossings across the built border occurred during the summer and early autumn of 2015, before EU member states had introduced temporary border controls. This allowed large groups of people to enter and exit territories without legal documentation once they had entered the EU. Similarly, the temporary lifting of borders could occur when border control was enforced only as spot checks, leading irregularised migrants to cross borders, for instance, on trains, or as a result of "fissures" in the border enforcement. An instability of the border emerges concurrently through our different materials. For border crossers, the border environment could change very quickly, for instance, due to weather conditions, border closures ahead, or border police approaching at certain times (cf. Mollerup 2020). This border instability requires attentiveness to the here and now of the border characteristic for *border navigated* as it emerges from our social media data through posts, such as the update on the Hungarian border situation presented above, and through efforts to obtain information, such as:

> [Social media data] Who has more information about the Danish border today? Posted on Arab Hamburg, 10 September 2015, translated from Arabic.

There is often a future orientation to these updates, which aim to facilitate or inform navigation. The border-crossing practices comprise navigation both *through* and *around* the built border. Turning to our ethnographic material, the affinity between traced lives and experienced lives in navigating the instability of the border becomes visible:

> [Ethnographic material] After arriving in Northern Europe in September 2015, Ghada and Ziad took a regular bus, trying to blend in among regularised travellers. As they reached the border, a police officer entered the bus and asked everyone to take out their passports. Ghada silently told her husband to remain quiet as the police officer approached them. She then explained to her that they were from Lebanon but living in Germany, and that they were on their way to visit her uncle who lived in the country they were trying to enter, a story which only bore a vague connection to the truth. The police officer asked if they had their passports with them. Ghada confirmed, hinting at her bag. Without asking them to show their passports, the police officer let them stay on the bus. Ghada and Ziad saw other irregularised migrants being led off other buses at the border, but they were themselves allowed to cross the border despite never showing their – non-existent – Lebanese passports. Ziad and Ghada, Syrian refugees, based on interviews on 3 and 20 March 2019.

The instability of the border is also followed closely by solidarians, who would assist irregularised migrants in crossing the border or negotiate with border authorities concerning irregularised migrants' rights and how to best manage the border in specific situations. This is characteristic for the *solidary border*. In the following excerpt from our social media data, we learn how the *solidary border* can simultaneously cooperate with and work against authorities. In the German–Danish borderlands, the Flensburg-based network Refugee Welcome Flensburg announced through a press release that they had decided to follow new measures after the Danish border police had forced migrants to register upon arriving in Denmark from Germany. Refugee Welcome Flensburg had previously assisted irregularised migrants by providing donation-funded tickets for the transit buses ("Bahn und Schienenersatzverkehr") provided by Deutsche Bahn due to overfilled trains and deliberately with the purpose of avoiding "Schleusserei," i.e., illegal border crossings:

> *[Social media data] Press Release: The organisational team of the initiative 'Refugees Welcome Flensburg' reacts to the current conduct of the Danish (border) police. This happens after two public transit buses [Bahn und Schienenersatzverkehr] within the last 24 hours, shortly after arrival at the Danish border, have been stopped in order to forcibly register all the refugees onboard [which is why] the volunteers feel incapable of assisting with further transit help. For the past 7 weeks, the volunteers at the Bahnhof provided humanitarian aid and thereby worked cooperatively and transparently together with all the relevant authorities on both sides of the German-Danish border.* Posted on Wir Sagen Moin (repost from Refugees Welcome Flensburg), 29 October 2015, translated from German.

Realising that the Danish border police would stop transit buses in order to register the irregularised migrants upon arrival, Refugees Welcome Flensburg ceased providing this kind of transit assistance:

> *[Social media data] So far, the refugees have been able to travel largely unhindered. The form of the random spot checks was typical of counterterrorism measures. The current controls can only be interpreted as deterrent measures. People's fates are thereby at risk. The current situation lacks any clarity, as the transports are stopped randomly, making it impossible for the organisational team to inform refugees about their onward journey. From the team's point of view, the assistance at the station can currently be limited to supplying only food and clothing.* (ibid.)

Similar kinds of stories are told in our ethnographic material. These materials move alongside one another without being directly linked to each other. In the following case, we highlight an example characteristic of the *politico-legal border*, which also focuses on borders as a matter of political negotiation and is thus addressed both at particular borders and in places well removed from these. For a volunteer based in Denmark, this border contestation played out when he learned from a secret Facebook solidarity group that a Rudolf Hess march was planned in Hamburg at the same time as the German–Danish border was being closed, leaving thousands of irregularised migrants stranded in Hamburg with thousands of neo-Nazis gathering. Realising the severity of the situation, he called the Danish police and explained, as he recalled us:

> *[Ethnographic material] If you close the border, there is going to be a large accumulation of refugees at the train station in Hamburg. Then there is no one who can protect these people against the Nazis (...). So, in the worst case, it will cost human lives, and there are certainly a lot of refugees who will get some unusually unpleasant experiences.* Jens, Danish solidarian, 21 March 2019, translated from Danish.

The Danish border did indeed remain open at this time. While our social media data would typically not reveal such detailed insights into conversations between solidarians and the police, this data shows various traces of similar negotiations with the authorities to the one mentioned here.

The contrapuntal move of bringing our different research materials into conversation does more than just show how experienced and traced lives are part of the same world. When brought together, they also document the instability of the border in ways that transcend the particular case. The unstable border relates to practices of *border navigated* as well as *solidary border* and *politico-legal border*. Finally, the contrapuntal move allows for new narratives of the border: the border's instability is indicative of its fragility can be contested, crossed, and conquered (cf. Hess and Kasparek 2017).

Contrapuntal Moves: Conclusion

In this contribution, we have rethought the relationship between social media data and ethnographic materials by establishing them as *contrapuntally* interconnected. With this contrapuntal move, we have actively sought to advance beyond logics of quantification and conventional scales of big and small, thick and thin, foreground and background that are offered in the computational data processing programs. For this purpose, we created a data environment suitable for transcending the criteria of relevance built into the software's logic (such as spikes based on number of entries). Pursuing Tsing's non-scalability approach, we refrained from nesting the "smaller" ethnographic material into the "larger" social media data.

Positioning our contribution in relation to similar, recent attempts at working with digital methods and ethnographic materials, we moved beyond stitching, complementing, assembling, or remixing strategies. Alongside Munk's four models for overcoming the qualitative–quantitative divide (complementarity, single-level, curating, and algorithmic sense-making), we positioned our research at the overlap between single-level and curating. We thus suggest that contrapuntal analysis could be thought of as a fifth model for overcoming the qualitative–quantitative divide.

By establishing relations in our material as contrapuntal, we have explored affinities between seemingly disparate materials: the social media data and our ethnographic material. We have done so with the overarching aim of identifying potentials and further questions for digital migration research: On the one hand, we aimed to create a conceptual framework, beyond the existing practices regarding online ethnography, in which ethnographic material and social media data can co-exist without being forced into a relational structure. On the other hand, we wished to show how scrutinising digital fields of navigation among vulnerable groups, such as irregularised migrants, poses specific ethical challenges and methodological issues that, if anything, become more prominent when different types of materials co-exist. We hope that this contribution can stimulate further critical reflection in the field of digital migration studies, particularly regarding the conceptual understanding of how big social data and ethnographic material can be brought into conversation.

Developing a contrapuntal lens enables us to show not only the multiplicity of the border, enacted through these three different-yet-related versions of the border, but also how these border versions are related beyond the particular case. We thus argue that *the border multiple* (Andersen and Sandberg 2012) is likewise enacted in social media communications, which can be counted, measured, and visualised into peaks while retaining their qualitative affordances. The opposite is true as well: recognition of how the border is enacted through social media data has more far-reaching potential when brought into conversation with the ethnographic material. The different materials thus present different-yet-related paths to knowledge of the multiplicity of borders. The contrapuntal approach, we argue, thereby establishes ground upon which we can expand our ethnographic insights without compromising them. As we have demonstrated, the insights are expanded because of their connectedness to the social media realm established and highlighted through our contrapuntal lens.

Our contrapuntal analysis has thus allowed us, following Said's call, to bring together discrete and internally coherent yet simultaneously co-existent and interacting experiences. In so doing, we have enabled new narratives of the border. The border multiple is at once fragile and robust: fragile because the border can be destabilised through the actions of solidarians and irregularised migrants, and robust because of governments' increasing militarisation and fortification of the border. We suggest that this conversation on materials foregrounds the interconnectedness between practices of irregularised border crossings—an interconnectedness that cannot remove or diminish the chaos, changeability, and insecurity connected to these practices but that enables new narratives of the border's fragility.

Acknowledgements This research forms part of the interdisciplinary research project: DIGINAUTS: Migrants' Digital Practices in/of the European Border Regime, funded by the VELUX Foundations 2018–2020. We wish to thank EthosLab, ITU, and TANTlab, AAU, for discussions involving our work and research methodologies, including our ambition to engage with social media data and ethnographic materials in new ways.

NOTES

1. Thanks to our colleague Tine Damsholt for reminding us about the metaphor of the contrapuntal.
2. According to Rozakou (2018, 200) the term *solidarian* draws on a specific, emic notion (*allilegyoi*) of standing up in solidarity (2018, 190). The term has occurred in different, specific historical conjunctures and has been still more widely used, also among solidarity initiatives for refugee and asylum-seeker relief in Greece. However, we find that while acknowledging its localised legacies, several of the general traits of this notion can be applied to other local contexts as well. For instance, its anti-authoritarian legacy (in Greece related to the post-dictatorship anarchist movements) as well as the

absence of, and in some cases deliberate distancing from humanitarianism discourse, applies to several of the solidarity initiatives included in our study.

3. We are indebted to Alaa Almeiza for transcribing our Arabic interviews into English translations.
4. All translations have been done by the authors. Thanks to Anna Sandberg, Annika Lindberg, and Mahmoud Alsayed for checking, respectively, German, Swedish, and Arabic translations.
5. All names are made up for the protection of research participants.

BIBLIOGRAPHY

Adamson, Fiona B., Priya Kumar. 2014. "Imagined Communities 2.0: Space and Place in Tamil, Sikh and Palestinian Online Identity Politics." Paper presented at the 55th annual meeting of the International Studies Association, Toronto, Canada, 26–29 March.

Andersen, Dorte J., and Marie Sandberg. 2012. "Introduction to the Border Multiple." In *The Border Multiple. The Practicing of Borders between Public Policy and Everyday Life in a Rescaling Europe*, edited by Andersen, Dorte J., Marie Sandberg and Martin Klatt, 1–19. Ashgate Border Regions Series.

Blok, Anders, and Pedersen, Morten A. 2014. "Complementary Social Science? Qualitative-quantitative Experiments in a Big Data World." *Big Data & Society* 1 (2), 205395171454390. https://doi.org/10.1177/205395171454 3908.

Blok, Anders, Hjalmar B. Carlsen, Tobias B. Jørgensen, Mette M. Madsen, Snorre Ralund, and Morten A. Pedersen. 2017. "Stitching Together the Heterogeneous Party: A Complementary Social Data Science Experiment." *Big Data & Society* 4 (2). https://doi.org/10.1177/2053951717736337.

Borkert, Maren, Karen E. Fisher and Eiad Yafi. 2018. "The Best, the Worst, and the Hardest to Find: How People, Mobiles, and Social Media Connect Migrants in (to) Europe." *Social Media+ Society* 4 (1). https://doi.org/10.1177/2056305118764428.

Bruns, Axel, Anja Bechmann, Jean Burgess, Andrew Chadwick, Lynn S. Clark, William H. Dutton, and M. Zimmer. 2018. "Facebook Shuts the Gate after the Horse Has Bolted, and Hurts Real Research in the Process." *Internet Policy Review* 25.

Boellstorff, Tom, and Bill Maurer, (eds.). 2015. *Data, Now Bigger and Better!* Chicago: Prickly Paradigm Press (marketed and distributed by University of Chicago Press).

Carter, Daniel. 2018. *Reimagining the Big Data assemblage*. Thousand Oaks: Sage.

Creswell, John W. 2014. *A Concise Introduction to Mixed Methods Research*. London: Sage.

Curran, Sara. 2013. "What's So Important about Music Education?" *Educational Research and Evaluation* 19 (1): 981–999. https://doi.org/10.1080/138 03611.2012.748249.

Diminescu, Dana. 2008. "The Connected Migrant: An Epistemological Manifesto." *Social Science Information* 47 (4): 565–579. https://doi.org/10. 1177/0539018408096447.

Gillespie, Marie, Souad Osseiran and Margie Cheesman. 2018. "Syrian Refugees and the Digital Passage to Europe: Smartphone Infrastructures and Affordances." *Social Media+ Society* 4 (1): 1–12. 2056305118764440.

Grosser, Benjamin. 2014. "What Do Metrics Want? How Quantification Prescribes Social Interaction on Facebook." *Computational Culture* 4. http://computationalculture.net/what-do-metrics-want/.

Haraway, Donna. 1988. "Situated Knowledges: The Science Question in Feminism and the Privilege of Partial Perspective." *Feminist Studies* 14 (3): 575–599.

Hess, Sabine, and Bernd Kasparek. 2017. "Under Control? Or Border (as) Conflict: Reflections on the European Border Regime." In *Perspectives on the European Border Regime: Mobilization, Contestation, and the Role of Civil Society*, edited by Ove Sutter and Eva Youkhama. *Inclusion*, vol. 5: 58–68.

Ingold, Tim. 2013. *Making: Anthropology, Archaeology, Art and Architecture.* New York: Routledge.

Ingold, Tim. 2018. "One World Anthropology." *HAU: Journal of Ethnographic Theory* 8 (1–2): 158–171. https://doi.org/10.1086/698315.

Kennedy, Helen, and Rosemary Lucy Hill. 2018. "The Feeling of Numbers: Emotions in Everyday Engagements with Data and Their Visualisation." *Sociology* 52 (4): 830–848.

Kitchin, Rob. 2014. "Big Data, New Epistemologies and Paradigm Shifts." *Big data & society* 1 (1): 2053951714528481.

Kok, Saskia, and Richard Rogers. 2017. "Rethinking Migration in the Digital Age: Transglocalization and the Somali Diaspora." *Global Networks* 17 (1): 23–46.

Kotsios, A., M. Magnani, D. Vega, L. Rossi, and I. Shklovski. 2019. "An Analysis of the Consequences of the General Data Protection Regulation on Social Network Research." *ACM Transactions on Social Computing* 2 (3): 1–22.

Komito, Lee. 2011. "Social Media and Migration: Virtual Community 2.0." *Journal of the American Society for Information Science and Technology* 62 (6): 1075–1086.

Latour, Bruno 2005. *Reassembling the Social: An Introduction to Actor-Network-Theory.* Oxford: Oxford University Press.

Latour, Bruno, Pablo Jensen, Tommaso Venturini, Sebastien Grauwig, and Dominique Bouiller. 2012. "'The Whole Is Always Smaller Than Its Parts'— A Digital Test of Gabriel Tardes' Monads." *The British Journal of Sociology* 63 (4): 590–615.

Law, John and John Urry. 2005. "Enacting the Social." *Economy and Society* 33 (3): 390–410.

Leurs, Koen, and Kevin Smets .2018. "Five Questions for Digital Migration Studies: Learning from Digital Connectivity and Forced Migration in(to) Europe." *Social Media and Society* 4 (1): 1–16. https://doi.org/10.1177/2056305118764425.

Leurs, Koen and M. Prabhakar. 2018. "Doing Digital Migration Studies: Methodological Considerations for an Emerging Research Focus." In *Qualitative research in European Migration Studies*, edited by Ricard Zapata-Barrero and Evren Yalaz, 247–266. IMISCOE Research Series. Cham: Springer. https://doi.org/10.1007/978-3-319-76861-8_14.

Lomborg, Stine, and Anja Bechmann. 2014. "Using APIs for Data Collection on Social Media." *The Information Society* 30 (4): 256–265.

Manovich, Lev. 2011. "Trending: The promises and the Challenges of Big Social Data." *Debates in the Digital Humanities* 2 (1): 460–475.

Markham, Annette. 2013. "Remix Cultures, Remix Methods." In *Global Dimensions of Qualitative Inquiry*, edited by Norman K. Denzin and Michael D. Giardina, 63–81. Walnut Creek: Left Coast Press.

Markham, Annette. 2017. "Remix as a Literacy for Future Anthropology Practice." In *Anthropologies and Futures. Researching Emerging and Uncertain Worlds*, edited by Juan Francisco Salazar, Sarah Pink, Andrew Irving, and Johannes Sjöberg, 225–241. London: Bloomsbury Academic. https://doi.org/10.4324/9781003084570-14.

Mol, Annemarie. 1999. "Ontological Politics. A Word and Some Questions." *The Sociological Review* 47 (51): 74–89. https://doi.org/10.1111/j.1467-954X.1999.tb03483.x.

Mollerup, Nina Grønlykke. 2020. "Perilous Navigation—Knowledge Making with and without Digital Practices during Irregularized Migration to Öresund." *Social Analysis—The International Journal of Anthropology* 64 (3): 95–112. https://doi.org/10.3167/sa.2020.640306.

Mollerup, Nina Grønlykke, and Marie Sandberg. Forthcoming. "'Fast Trusting' – Practices of Trust during Irregularised Journeys to and through Europe." In *The Migration Mobile: Border Dissidence, Sociotechnical Resistance and the Construction of Irregularised Migrants*, edited by Vasilis Galis, Martin Bak Jørgensen, and Marie Sandberg. Lanham: Rowman & Littlefield.

Munk, Anders. K. 2019. "Four Styles of Qualitative-Quantitative Analysis: Making Sense of the New Nordic Food Movement on the Web." *Nordicom Review* 40 (1): 159–176.

Rogers, Richard. 2013. *Digital Methods*. Cambridge: MIT Press.
Rogers, Richard. 2018. "Otherwise Engaged: Social Media from Vanity Metrics to Critical Analysis." *International Journal of Communication* 12: 450–472.
Rozakou, Katerina 2018: "Solidarians in the land of Xenios Zeus: Migrant Deportability and the Radicalisation of Solidarity." In *Critical Times in Greece: Anthropological Engagements with the Crisis*, edited by Dimitris Dalakoglou and Georgios Agelopolouse. Oxfordshire: Routledge.
Said, Edward W. 1994. *Culture and Imperialism*. New York: Vintage Books.
Said, Edward. 2000. "Reflections on Exile." In *Reflections on Exile and Other Essays*, 173–86. Cambridge: Harvard University Press.
Sandberg, Marie. 2016: "Restructuring Locality: Practice, Identity and Place-Making on the German-Polish Border." *Identities—Global Studies in Culture and Power* 23 (1): 66–83.
Sandberg, Marie. 2020: "Retrospective Ethnographies: Twisting Moments of Researching Commemorative Practices among Volunteers after the Refugee Arrivals to Europe 2015." In *Challenges and Solutions in Ethnographic Research: Ethnography with a Twist*, edited by Tuuli Lähdesmäki, Eerika Koskinen-Koivisto and Viktorija L. A. Čeginskas, 117–130. London: Routledge. https://doi.org/10.4324/9780429355608.
Snelson, C. L. 2016. "Qualitative and Mixed Methods Social Media Research: A Review of the Literature." *International Journal of Qualitative Methods* 15 (1): 1609406915624574.
Strathern, Marilyn. 1991/2004: *Partial Connections*. Updated edition 2004. USA: Altamira Press, Rowman & Littlefield.
Tsing, Anna. 2012. "On Nonscalability: The Living World Is Not Amenable to Precision-Nested Scales." *Common Knowledge* 18 (3): 505–524.
Tufekci, Zeynep. 2014. "Big Questions for Social Media Big Data: Representativeness, Validity and Other Methodological Pitfalls." In *Proceedings of the International AAAI Conference on Web and Social Media*, vol. 8, no. 1.
Wang, Tricia. 2013. "Big Data Needs Thick Data" [blog post]. Retrieved from http://ethnographymatters.net/blog/2013/05/13/big-data-needs-thick-data/. Last accessed March 12, 2021.

CHAPTER 4

Migration Trail: Exploring the Interplay Between Data visualisation, Cartography and Fiction

Giacomo Toffano and Kevin Smets

Introduction

This chapter outlines recent evolutions in data visualisation of migration and investigates the possibilities for its blending with cartography and fiction. The study orbits around Migration Trail, an online interactive platform that allows users to follow a real-time journey of two imaginary migrants to Europe through the use of maps, podcasts, fictional family conversations and interactive data visualisation. The project was created by Killing Architects, a Rotterdam-based independent architect studio run by Alison Killing. The studio experiments with communicating to the

G. Toffano (✉)
Brussel, Belgium
e-mail: giacomo.toffano@vub.be

K. Smets
Department of Communication Studies, Vrije Universiteit
Brussel, Brussel, Belgium
e-mail: kevin.smets@vub.be

M. Sandberg et al. (eds.), *Research Methodologies and Ethical Challenges in Digital Migration Studies*, Approaches to Social Inequality and Difference, https://doi.org/10.1007/978-3-030-81226-3_4

general public about social issues that have a strong spatial component such as migration and climate change, exceeding the conventional architectural production of design and plans, to create exhibitions, writing, film and research. In Killing's words, her studio attempts to take distance from a traditional interpretation of her profession to harness "architects' skills of spatial analysis and representation" beyond the mere construction of buildings (Killing 2018, 32). The authors have opted for an in-depth case-study approach as the project is a highly original and novel effort to challenge the way in which migration data are processed, combined and communicated. Considering the interplay of geospatial data visualisation with other representational techniques, the research examines how it challenges the ubiquitous narrative of migrants as problematic and vulnerable "others". With the use of multimodal analysis, this chapter investigates the interaction of textual, audio, visual and spatial elements of communication in Migration Trail.

In stark contrast to the "invasion of red arrows" often observed in traditional EU migration cartography (Van Houtum and Bueno Lacy 2020, 201), this website renders an unusual set of data (i.e. wind strength, airplane routes, battery level of the migrants' mobile phones and GPS coordinates of previous migrants' shipwrecks) along with fictional family chats, and original atlases of human mobility. Migration Trail takes advantage of recent developments in digital communication technologies, specifically by benefiting from the proliferation of affordable web-based geographic information systems that allow independent content producers to manipulate and display great volumes of geographic data. Once again—as envisaged in the introduction of this book—digital technologies reshape the way producers tell migration stories: autonomous users can instantly query (big) data to design interactive cartographies and visualisations, a digital immediacy that is prompting a genuine boom in neo-cartographic practices. Migration Trail's geospatial display is considered in all its individual elements, with each visualisation choice studied both in its singularity and in interplay with other communicative dimensions.

The chapter begins with a concise excursus on relevant literature exploring recent studies on migrants' cartographies, data visualisations and fictional accounts; three different fields that are proficiently intersected by Migration Trail's composite production. Subsequent sections outline in detail how the website functions and summarise the results of the analysis. All in all, the multimodal exploration highlights how Migration Trail attempts to clearly illustrate the humanity of migrants by vividly narrating a subjective account of human mobility. The remarkable use of hybrid visualisations also emphasises the structural—political, economic and social—dimensions of migration in an attempt to subvert hegemonic mapping practices by challenging the ever-present EU border securitisation perspective. However, the conclusive analysis of Migration Trail's narrative strategies highlights certain problematics, tensions, as well as several limitations in the representation of migrants, which could possibly hamper the website's effort to convey an original mobility story through the interplay of fiction, cartography and data.

Migration Data Visualisations, Cartographies and Fiction

In the words of its creator, Migration Trail is a fictional online experience narrated through a "real-time animated mapped data visualization" (Killing 2018, 33). A significant portion of the project, indeed, consists of visual representations of spatial data, originally assembled to accompany the narration of two migrants in their journey to Europe. A cartographical view evolves over the course of 10 days from the first opening of the website and serves as the background for the interactive data visualisations and narration of the fictional border crossings. The project addresses a non-expert audience, as Killing presents the project as stemming from the recognition that migration is typically poorly understood among European news audiences. The creator criticizes the fragmented mainstream media news coverage that seldomly goes beyond reporting individual tragic events, restraining the general public's ability to develop a systemic understanding of migration. This section explores recent literature on each of the three elements interacting in Migration Trail—cartographies, data visualisations and fictional accounts—setting the scene for the following detailed scrutiny of their interplay in the overall experience.

Our research embraces the call for a critical understanding of migrant representation in data visualisations (Crawford et al. 2014, 8; Ruppert et al. 2017, 2). In this light, all data artefacts should be considered as multimodal objects that are both "generated by – and generative of – data politics" (Allen 2020, 186). Specifically, data visualisations—maps, tables or charts—intervene in political debates, revealing or challenging established discourses, norms and hierarchies of values (Ruppert et al. 2017, 2).

Datasets might reproduce choices by visual designers and commissioning institutions, reflecting graphical constraints but also echoing political priorities and policy-making agendas.

Scholars investigate the effects of specific design decisions to communicate data on human mobility and examine how compositional choices—of written, visual and spatial elements—can convey both express and tacit political positions about migration (Allen 2020, 180–191; Rall et al. 2016, 171–197; Risam 2019, 566–580). Risam notes how migration data visualisations frequently corroborate problematic framings of migrants: dehumanising individuals and marginalising vulnerable communities with visual discursive choices (ibid.). In the same spirit, Rall reviews several data visualisation choices, advancing strategies to strengthen data communications for human rights advocacy and avoids the above-mentioned adverse effects of aseptic datasets on migration (Rall et al. 2016). Over the past three decades, a plethora of critical approaches have highlighted how mapping practices—like any other data visualisation—can reproduce cultural power relations: far from considering maps as mere objective reflections of reality, scholars have focused on arbitrary choices, frames and discourses that often subject cartography to the hegemonic power of political authorities (Crampton 2009, 91–100; Harley and Laxton 2002; Monmonier 2018). In this respect, researchers have noted how human mobility maps often reflect certain cartographical "distortions", possibly affecting the public debates on borders and undocumented migration across the Union (Van Houtum and Bueno Lacy 2020, 201).

Inspired by such critical approaches, recent papers on migration cartography have attempted to capture original efforts that aimed to go beyond the hegemonic use of mapping: initiatives that reclaimed the potential of maps to contest the normative geographies of mobility that are inborn in the practice (Tazzioli 2019, 397–409). To illustrate this approach, Mekdjian and Amilhat Szary explore "counter-cartographies of exile" by studying a map that, beyond countries and cities, includes personal traces of the refugees' experiences on the road to Europe (i.e. governmental and material restrictions, the challenges of clandestine truck rides, the risk of police encounters) (Mekdjian and Szary 2018, 258–263). Similarly, Lo Presti analyses EXODI, an example of interactive cartographic narration that includes the first-person experience of thousands of migrants from Sub-Saharan Africa (Lo Presti 2020, 911–929).

Van Houtum and Bueno Lacy suggest a useful typology of such alternative cartographies, distinguishing three types of developing practices:

counter-mapping, deep mapping and mobile-mapping (Van Houtum and Bueno Lacy 2020, 201). The examples in the first category (e.g. Topological Atlas)[1] aim at countering "the languages and images of power" to become means of resistance and emancipation (Nancy Lee Peluso 1995, 386). Similarly, deep mapping counters the polished, flat representations of hegemonic maps through the portrayal of subjective accounts (e.g. Migreurop's Mustafa's Journey) (Migreurop 2017). The third strand, mobile mapping uses mobile-data "phones, social media, cameras, satellites, open-source mapping and film" to render the highly subjective, complex and polymorphous nature of human mobility (Van Houtum and Bueno Lacy 2020, 16).

Migration cartography, as mentioned, has evolved beyond the mere geolocation of mobility tracks to encompass seizing maps' virtually unlimited potential to decipher and tell stories (Caquard and Cartwright 2014, 101–106). In this regard, combining fiction with maps—and other types of data visualisations—represents a compelling development in the field of experimental, alternative representation practices. Original works in this terrain (i.e. 407 Camps, Schaal's Cartography, or Crossing Maps)[2] remain largely unexplored in academic research to this day.

Giada Peterle's conceptualisation of carto-fiction as a "self-reflexive, ethnofictional, creative carto-centred practice" represents a notable exception in the academic panorama (Peterle 2019, 1070–1093). The author presents a concise novel *Unfolding Berlin*: a production that uses words, maps and drawings to narrate the personal exploration of an unknown, unfamiliar city. The explicit aim of such a hybrid effort is to render an emotional, intimate cartography, producing a narrativising of mapping practices from a post-representational perspective (ibid., 1074).

Other than that, studies on fictional migration narratives have limited their explorations to traditional domains such as film, documentaries and literary fiction. Over the past two decades, migration fiction in both films and literature has developed at a fast pace and is now considered a self-standing genre (Mandelbaum and Ridet 2011). Academics have investigated the descriptive constructions of the genre, identifying recurrent frames and discourses in that field (Berghahn 2020, 399–410; Hiltunen 2019, 141–155; Ponzanesi 2016, 217–233; De Bruyn 2020, 25). Such studies delved into migration documentaries, films and books, producing significant reflections on concepts such as kinship, bordering, dehumanisation and othering. This has put on the map efforts to portray migrants' subjective experiences, and intimate realities narrated in migration fiction. Other investigations on the same topics alternatively noted how in artistic discourses on refugees and migration, creators often reproduce stereotypical identities inherited from mainstream media (Drüeke et al. 2021, 160–183).

All in all, research on data visualisation, cartography and fiction emphasises a certain level of tension between accounts that reproduce hegemonic discourses about migration and alternative efforts to produce cultural artefacts that counter such narratives. On the one hand, discourses that are widespread in traditional press—the ubiquitous portrayal of migration as a "crisis", and of migrants as suffering or dangerous others (Berry et al. 2016; Chouliaraki et al. 2017, 35)— get picked up, validated and reinforced by certain uncritical digital media productions (Aarssen 2017, 1–14). On the other, researchers have directed their attention to grassroots digital practices, exploring efforts to counter such persistent narratives (Nikunen 2020, 411–423; Georgiou 2018, 45–57). In several online practices, researchers have found personalised accounts of migrants: individuals who can "voice" their experience and appear as campaigners, militants and political agents (Georgiou 2018, 58). Their suffering is, in such communicative efforts, profoundly de-massified, shedding light on stories of individual struggle and hope. Inspired by Georgiou and Nikunen's explorations, this paper intends to shed light on Migration Trail by exploring how its unusual combination of visualisation practices provides migrants with "voices" and counters widespread stereotypical discourses on migration. Since 2015, a whole body of counter-cartographies of migration emerged on the media landscape. Among such efforts, this chapter focuses specifically on Migration Trail because of its highly intermedial character, intended as the combinations of several modalities of interaction with the audience. Such pronounced intermediality denotes Migration Trail as a highly hybrid media product that simultaneously intersects three different scholarly debates on forced migration. Primarily, Migration Trail's narration through a textual message box recalls the scholarship of the digitally connected migrants. Secondly, the interactive mapping experience prompts us to reflect on the increasing possibility of proficiently using digital tools for migration storytellers. Finally, the choice of creating fictional characters that venture into a migrant journey intersects the debates on the truthful-objective migration accounts and creative representation of minorities in the media.

RESEARCH OBJECTIVES OF THE ANALYSIS

This study considers Migration Trail's interplay of data visualisation, cartography and fiction, examining to what extent such a hybrid product challenges the hegemonic portrayal of migration as a "crisis", and of migrants as suffering or dangerous others. Alison Killing created Migration Trail intending to mend for allegedly flawed media coverage of migration. Our analysis of the website descended from such consideration and began with the exploration of academic literature on mainstream media accounts of migrants (e.g. Chouliaraki et al. 2017). The portrayal of migrants as *others* to European audiences emerged as the ubiquitous critique scholars moved to mainstream media accounts, providing a relevant conceptual lens to empower the exploration of Migration Trail.

The concept of othering, central to the present exploration, is considered a dichotomous recognition of identities that is instrumental in the process of exclusion of specific individuals from European societies (Udah and Singh 2019, 843–859; Zaborowski and Georgiou 2019, 92–108). Representations that slip in depicting migrants as others, often reproduce them in fallacious, dichotomous symbolism: the speechless victim and the evildoing culprit (Chouliaraki et al. 2017). In common press accounts, in traditional data visualisations or fiction, migrants often appear as pure victims: passive subjects of political-economic conditions and vulnerable beings in need of protection. Yet, at other times, they emerge as a threat to European societies, or a menace to the cultural, organisational and social welfare of European countries.

This chapter particularly concerns deepening an original narration of ordinary experiences of mobility. To this end, we cast academic light on Migration Trail's attempts to describe migrants' connected lives, affective interactions and networks of care, and therefore respond to the autonomy of migration (AoM) call to overturn the scholarly fixation with governance and security. Conversely, such a scholarship invites engaged academics to replace governance and security by investigating opportunities to create a "common world of existence": highlighting migrants' agency to occupy social spaces (Papadopoulos and Tsianos 2013, 178).

This research, therefore, aims at understanding to what extent Migration Trail fulfils its commitment to produce a militant, original portrayal of the everyday sociability of mobile individuals. In particular, the study deepens understanding of how the introduction of a fictional narrative—portrayed through an unusual set of data visualisations and maps—challenges the dominant othering of migrants. Empirically, Migration Trail's content, both textual and visual, is explored in relation to the aforementioned discursive categories of vulnerable and dangerous others, assessing the author's effort to counter those two frames emerging prominently in traditional data visualisation, cartography or fiction.

Initially, this research attempts to map Migration Trail's functions, thoroughly exploring the website's content while understanding the dynamic interplay of all multimedia elements that interact in the digital visualisation. The objective is to detail how the website operates, what techniques are employed and how verbal, visual, textual and sonic elements work together in the communicative effort of this highly hybrid web piece. Subsequently, the analysis focuses on how Migration Trails renders its unconventional narrative about migration. The effort lies in uncovering what messages and frames the creator embedded in the website, identifying how migration is discursively conceived and delivered, and verifying the extent to which such web work challenges the othering of migrants perpetrated by a significant part of the media.

The study starts out with a multi-method approach to the content, combining multimodal analysis (Pauwels 2012, 247–265), which is further validated by a semi-structured interview (Adams 2015, 492–505). With multimodal analysis, the authors investigated the website in all its components, focusing on the interactions and combinations of different modes of communication to achieve a communicative function (O'Halloran and Smith 2012). In other words, the analysis considers the interplay between the multiple modes (i.e. texts, maps, videos, sounds or music) engaged in creating the website. The choice of a multimodal framework of investigation implied an iterative process during which components and layers of the website were progressively analysed in more detail. This method resonates well with the highly interactive nature of experimental data visualisations on human mobility, i.e. productions in which the medium has a particular influence on the choice of specific modes of expression (Hiippala 2020, 277–291).

In stage two of the study, the authors shared the findings of the multimodal analysis with Alison Killing, the creator of Migration Trail. This inquiry followed the scheme of member checking: a well-established practice used to counter-check pre-analysed data to integrate and confirm the validity of such information (Cho and Trent 2006, 319–340). Through an interview, the authors shared the inferences, beliefs and emotions gathered throughout the analysis, aiming to see how such results accurately render Killing's intentions when producing the website. This study attempts to generate what Geertz (1973) described as a "thick description" of Migration Trail. In this light, we foresee an interaction between the multimodal analysis and a semi-structured interview with the creator as a particularly suitable mixed method to analyse the data. However, the research is grounded in an interpretive research field, and there is no purpose in seeking an absolute correspondence with the realities investigated. Therefore, the multimodal analysis and the interview should be seen as decoupled with the aim of truthful validation; on the contrary, they are solely intended to integrate each other to the overall goal of providing a particularly detailed account of the website analysed.

Denotative Investigation of Migration Trail: An Overview of the Website

This section offers a comprehensive descriptive account of Migration Trail and outlines in detail the three main sections of the website: the cartographical view, the fictional chat box and the spatial data visualisations.

The Main Cartographical View

From a denotative perspective, Migration Trail's main visualisation presents the user with a full-screen cartographical view, a visualisation that is popularised by digital delivery and mobility services such as Deliveroo and Uber. As noted by McKinnon, the advancements in the digital domain enable the proliferation of maps, which has led researchers to declare that we live in the age of mapping (McKinnon 2020). Web-based geographic services (such as Mapbox and Google Maps) have simplified the production of cartographies, and now creators can, within the limits of the technological affordances, appropriate and remodel online maps for their own purposes. Migration Trail's opening map is initially centred in Libya's capital, Tripoli, with a large zoom feature giving a broad view of the Sicilian Channel. This geographic segment was rendered ubiquitous in mainstream media's representation of the extended European maritime surveillance apparatus, and this focus is reminiscent of the monitors and radar of a European border patrol station. Such aerial displays–as noted by

Madörin (2020, 698–711)—often risk producing the effect of commodifying refugees and reducing their movement to statistical probabilities. The website's data visualizations, maps, chatbox and sounds evolve over ten days from the moment the user first opens the page, inviting the audience to repeatedly access Migration Trail to obtain a protracted and continuous engagement with the narration. The connection between written and visuals elements is central to the overall 10-day experience in Migration Trail. As is often the case, visuals and written content coexist in a highly complementary context, jointly providing new meaning to the whole construction (Martinec 2005, 343).

The two human protagonists, identified on the map by pulsating red dots, move around the cartographical visualisation while their fictional conversations begin to pop up in the chat box. The main cross-modal correlation lies, therefore, between the visualised position on the maps and the textual content: the interplay between the two elements provides the full picture of the journey narrated. Meanwhile, background elements illustrate the context of action: wind visual signifiers, previous shipwrecks icons, stylised planes and ships moving on the maps play evocative symbolic roles in relation to the characters' journeys. An additional cross-modal correlation lies between sound and visuals, as Migration Trail's sonic signifiers feature prominently in the overall visualisation, fulfilling an expressive-symbolic purpose (Chion and Gorbman 2019). The soundtrack follows the protagonist's mood and condition, conveying feelings that are contextual to the situation portrayed in action. Rather than providing realism to the creation, such background music seems intended to increase the empathic dimension conveyed by the interactive map.

The Fictional Chat Box

The bulk of Migration Trails' fictional content resides in the visualisation of the chat box (a standard messaging interface popularised by services such as Telegram and WhatsApp). The conversations—written beforehand by the author—appear progressively on the screen, reporting messages sent by two migrants to their respective relatives. The messages pop up in the chat box over the course of ten days, and the text flow is unidirectional, featuring only messages composed by the two fictional protagonists.

The male character, David Ighiwiyisi, delineates an ambitious Nigerian man with a prominent entrepreneurial attitude. His business aspirations appear as the straightforward, main motives prompting his trip through Libya, Italy and France. In parallel, his messages display strong networks of family relations and fraternal bonds that tie him to his home country.

The second fictional protagonist, Sarah, a 19-year-old Syrian, exchanges messages with her brother while proceeding through the "Balkan Route". As her red dot moves on the cartographical visualisation, she fictionally crosses Turkey, Greece and Germany, constantly updating her brother on her whereabouts. The introduction of a female character provides an often-disregarded gender perspective: traditional coverage of Syrian asylum seekers prioritised visual representations of large groups of men, images that were fed intentionally or unintentionally into the trend of portraying foreign others framed as security or economic threats (Amores et al. 2020, 291–314).

Extracts from Sarah's chat box

Omar! No not yet, we're still at home, Mum managed to make us some koussa mehshi for our final meal, going to miss this so much! ... Omar I can't stop visualizing their faces, I'm so sure Dad was holding back tears!

Hi, fell asleep! Soooo Sorry. Promise!

I need to thank you for getting me through the day. I love you so much Omar. You gave me so much strength.

Throughout the 10-day experience, David and Sarah report on their journeys to Europe. Through the vehement exchanges with their relatives, the migrants describe violent police raids and threatening encounters with hostile individuals. At other times, they tell of particularly generous deeds: spontaneous acts of individuals and NGOs that help during the journey. The fictional chats allow Migration Trail to delineate the complex relationships between migrants and their smugglers, possibly going beyond the stereotypical portrayal of unambiguous exploitation. If the migrants' contempt towards reckless smugglers features heavily in their initial exchanges, throughout the days, they recognise how smugglers, notwithstanding their costly services, are key enablers for their movement.

Extracts from David's chat box

So, our guy Samir is connected to Mo who is the Libyan guy in charge of the boats... We can pay less if we want to go on a boat without GPS. They both have to make profit so they need enough people on a boat to make it. [...]

He doesn't have a smuggler so he cannot b smuggled into d goods yard where they can put u inside goods crossing the channel.

The text exchanges appear both on the website and through instant notifications on Facebook, until the end of the tenth day of the journey. Closing banners signal the conclusion of the experience, informing the users that Sarah has reached her final destination in Berlin and can successfully proceed to claim asylum, whereas David remains stranded in Calais, having failed previous attempts to reach the UK.

Spatial Data Visualisations

Along the 10-day virtual trail to Europe, the website progressively unlocks access to 18 spatial data visualisations: different maps that explore datasets displayed along national borders. Each tab—combining spatial statistics and a text frame—answers a particular question on migration, with widespread use of colour saturation and gradient. As van Leeuwen suggests, colour is frequently used in data visualisation to express emotive significance (van Leeuwen 2011, 563). In such displays, red is frequently connected with inaccessibility and isolation, whereas green often represents wide-ranging freedom of movement. Migration Trail exploits the communicative potential of the gradient and saturation to express intensity, with darker red countries granting their citizens the least degree of international mobility. For instance, in the dataset "Who needs a visa to visit the Schengen zone?", states are colour-coded to show the potential for their citizens to enter the Schengen area. Countries with no visa required are bright green, whereas countries whose citizens must fulfil visa requirements range from light pink to dark red, depending on the chances of receiving permission. Analogous colour schemes occur in each of the 18 cartographical views.

A rectangular switch allows users to shift among the different sets of geographical data. Users can quickly interchange between visualisations such as "Where can Syrians go without a visa?" and "World GNI per Capita", whereas the variations between the datasets are reflected in the countries' varying levels of colour saturation. Unsurprisingly, some visualisations (i.e. "Unemployment Rate around the World" and "Which are the Best and Worst Passports to Hold") show correlations and overlaps, with rough matches occurring throughout the 18 visualisations. Repetitive tones demonstrate patterns of nations that tend to lag behind in many indicators (GDP per capita, employment...), thus powerfully shedding light on enduring patterns of global inequalities.

Each dataset links to a different Migration Trail podcast, a ten-episode broadcast that features extensive contributions by officials, activists and migration scholars: complementing the fictional narration of the two trails with migration professionals' authoritative standpoints. Overall, this extensive collection of interviews attempts to give a systematic account of the structural conditions of migration. The underlying narrative clarifies how migration is an enduring phenomenon that is set to persist because of conditions of global destitution and social inequalities, but also because of shifting aspirations arising from development (Van Heelsum 2016, 1301–1309).

The Ethical, Political and Methodological Concerns of Migration Trail's Fictional Account

After thoroughly describing the website's elements, in this section we discuss their dynamic interplay, scrutinising the interactions of all multimedia elements that contribute to the website narrative. Migration Trail's main written communication consists of the two migrants' fictional conversations with their relatives: the protracted exchange of short texts is visually rendered through a live instant messaging feed that appears on the website chat box. Caine et al.'s work helps viewers to appreciate such unusual rhetorical choices for a cartographical effort. In their understanding, the use of fiction can create a new layer of empathy to deepen consciousness and provide a unique way for creators and participants to understand experiences (Caine et al. 2017, 215–221). Fiction, therefore, might allow a different kind of reaching out, contributing to the creation of a more humane connection with the audience. Designing a fictionalised world might vividly tighten the relationship between creators and participants, exposing further imagination and playfulness. By spreading the transmission of the fictional texts over 10 consecutive days, Migration Trail distributes the users' engagement over time, possibly prompting a longer sedimentation of the meaning along with deeper knowledge reflection and consumption.

Beyond its communicative potential, the use of fiction on the website certainly responds to the need to conceal vulnerable individuals, protecting them from dangerous public exposure. The website displays a disclaimer informing viewers that "the characters in the data visualization are fictional but based on true stories. Some locations relevant to the story have been concealed in order to protect vulnerable people".[3] This sentence connects the use of fiction with the need to address ethical concerns. Clandinin (2013), among others, notes how fiction can serve to shield participants from being identifiable, especially when they disclose tension-filled accounts of personal experiences (ibid.). For Claudia Mills, the power of sharing real-life accounts should always be evaluated against the potential dangers of embarrassing or betraying others (Mills 2004, 101–120). The writer or creator needs to behave as a responsible moral agent who seeks to reduce possible harm (ibid.).

Simultaneously, Cosgrove notes how, at times, fiction writers might have "escaped scrutiny in the ethics debate because the subjects under analysis — characters — are not real" (Cosgrove 2009, 1–134). In response to this, the same author proposes that writers should consider ethics of representation when discussing the narrative aspects of fictional stories. This also extends beyond traditional writing, to creators of fictional narrative content. Especially because of the engaged tone of Migration Trail, the ethical repercussions of its fictive narrative must be carefully evaluated. Such a critique should not disregard that, to successfully challenge existing power relations, fiction must go beyond archetypal portrayal of vulnerable groups: prompting the audience to recognise that their understanding might not be exhaustive and dissuading them from more explicit-meaning correlations and judgements (Chouliaraki 2006). As suggested in the following sections, a thoughtful evaluation of Migration Trail's narrative intent highlights important shortcomings that may possibly hamper its genuine idealistic commitment.

A first level of concern certainly connects with the ethical dimension of engaging with fiction to promote "a better-informed discussion about migration to Europe" (Killing 2018, 33). On the one hand, the widespread belief that fiction writers are given "poetic licence to break ordinary moral rules in the service of artistic creation" might induce scholars to neglect the ethical examination of fictional accounts because of their imaginary, made-up content (Mills 2000, 195). On the other hand, insurgent writers—those committed to improving public discussions about migration—should reasonably use such "poetic licence" to actively promote a new ethics of representation: using the fictional, imaginary potential of storytelling to show what is normally hidden from the public. In Migration Trail, the representation of two characters is regrettably not endowed with less typical profiles that could possibly have promoted better-informed discussions about migration to Europe.

Tellingly, Migration Trail's David is portrayed as the quintessential economic migrant: particularly in relation to his reasons for embarking on the journey, his character retraces certain elements of clichéd stories on his quest to individual success that corresponds well with traditional European or Western imaginaries. David's "dream of starting a business" and joining his brother, who reached England as "a budding footballer", is reminiscent of certain traits of migrants' caricatural descriptions of individuals as those ready to "get rich or die trying" (Shrestha 2020, 1–27). Similarly, Sarah represents a young Syrian woman with excellent grades at school, craving to enrol in a German university to "study Law or Political Science". Again, visibility seems to be granted to migrants' profiles that resonate well with Eurocentric ideas of deservingness: "dreams of individual success, hard-working ethos" (Georgiou 2018, 33).

A second problem regarding the representation of migrants concerns the scarce political agency attributed to the two protagonists. Throughout the fictional narration, the two protagonists refrain from making overtly political statements and never question the border policies that systematically dismiss their citizenship rights. As Killing admits, "they are just people busy living": they seem to proceed through their dramatic quest with resignation, without challenging the governmental dimension of their exclusion. To a certain extent, Migration Trail's accounts appear to support frames of vulnerability and threat, failing to question traditional representations of undeserving "illegals" seeking access to the European continent. Such accounts give travelling and mobility agency without offering political agency and might produce an unwanted naturalisation of migrant's conditional existence in Europe (Georgiou 2018).

As previously explained, the fictional chat box tells the story of two individuals introduced as a male economic migrant from Nigeria, and a female asylum seeker from Syria. The 10 narrated days culminate in a successful ending for the Syrian woman, who manages to reach Germany, whereas the Nigerian economic migrant remains stuck in Calais with few future prospects. Such highly dichotomous portrayals therefore risk replicating governmental divisions of individuals as deserving/undeserving of EU access and assistance: a framing that Migration Trail's creator wanted to challenge by working with "writers from Nigeria, Lebanon and Afghanistan to develop these characters and to write the migrants' voices" (Killing 2018, 33). As Tacchi notes, successful politics of recognition necessitate deep transformations in the orderly structure of voices and attention conferred on vulnerable actors and communities (Tacchi 2012, 225–241). This third problematic aspect of Migration Trail's narration therefore involves such attributions of voice: the decisions to identify and empower who gets to speak. Despite consulting an international pool of writers and ostensibly giving voices to two migrants, Migration Trail's fictional representations appear as yet another attempt to talk about migrants rather than letting them speak. Although such fictional engagement proficiently attempts to personalise migrants' faith and render their humanity, it also fails to provide a solid alternative to the Eurocentric standpoint of conceiving migration as a "problem" requiring authoritative and unequivocal remedies. Once again, the recognition of one more space of appearance does not seem to correspond with a full-fledged attribution of voice and an exhaustive turn of narrative point of view. In these regards, other militant-engaged initiatives substantiate how a fruitful collaborative process of co-creation can help address the many ethical, political and methodological concerns posed by the contemporary media productions regarding migration (e.g. Mekdjian and Szary 2018; Nikunen 2020). For instance, in one such project, Mekdjian's "Crossing Maps", the autoethnographic research involving a collaboration of migrants, academics and storytellers seems to pave the way for more coherent reflections on narratives about migration that distance themselves from stereotypical portrayals.

These three problematic dimensions could, to some degree, hamper the project goal to experiment with "how to create a better-informed public discussion about migration to Europe" (Killing 2018, 33), and reproduce some of the most typical shortfalls of traditional data visualisation, cartography and fiction about migration (Risam 2019, 566–580). When questioned on the topic, Killing appeared aware of the shortcomings and attributed them to the limited resources granted to the initiative. The initial idea included showing ten characters instead of two, a factor that probably would have allowed for more diverse accounts:

multiple stories that, indeed, could have more clearly highlighted the many complexities of human mobility.

Beyond the thorny issues just signalled, the website's driving narrative certainly constitutes an effort to re-humanise migrants' representations. The creator explicitly employs the fictional story to plunge the user into the private sphere of the migrants, exposing viewers to the individuals' family relations, and uncovering their aspirations, fears and hopes. The narration underlines the two characters, detailing their industrious attitudes, kinships and social capabilities, along with their personal expectations and fears. The two personal accounts appear, in these regards, as manifest efforts to counter the symbolic duality of the figure of the migrant, who, in traditional media, oscillates between the voiceless victim and the threatening other (Chouliaraki et al. 2017, 1–35). To re-humanise the migrants' quest, the main cartographical view is infused with graphical dynamic elements that highlight the epic dimension of migratory journeys, resisting the narrative of migrants as problems or mere statistical units (Madörin 2020, 698–711). Migration Trail makes full use of the possibilities of the digital medium to display a vast amount of interactive geospatial data: wind strength, battery level of the migrants' mobile phones, GPS coordinates of previous migrants' shipwrecks, along with other animated symbols that cast light on the everyday experiences of those on the move.

Challenging Traditional Media Portrayals?

Besides the threefold problematic of migrants' representations in the fictional chat box, several other elements of the website are more successful in challenging traditional media portrayals of migration. Visual representational signifiers are key for contextualising David and Sarah's trails on the main map: a multiplicity of dynamic elements, indicators such as wind speed and direction, aircraft routes and the data visualisation of the "power" of a Nigerian passport, present the user with a paradoxical contrast. They signal the abundance of legal, secure connections between European States, but also accentuate the reduced "capabilities" of the two protagonists embarking on such a hazardous journey.

Eighteen spatial data visualisations subvert hegemonic mapping practices, assuming the migrants' subjective point of view rather than the more common "EU border securitisation" perspective. In Migration Trail, colours highlight the reduced citizens' rights that migrants possess as travellers, reversing traditional map colour coding that often uses gradients and bold arrows to highlight the urgency of the migratory crisis. From a Nigerian citizen's viewpoint, most of the world is dark red: their passport is of little to no value, as it does not grant access rights to many countries. Such data atlases emphasise illegal migrants' *bare life*, subjects that, in Agamben's understanding, are alive but exist under circumstances with stripped-down citizenship rights (Agamben 1998).

Data visualisations communicate a powerful message, addressing structural and historical conditions underlying human mobility: visualisations that attempt to counter the narration of migration as an "unprecedented crisis", a contested framing that signals abrupt and disturbing events (Krzyżanowski et al. 2018, 1–14; Hiltunen 2019, 141–155). Such "crisis" discourses divert attention away from complicated, structural and historical developments of human mobility, a phenomenon that arguably has never ceased throughout history and encompassed a much more extensive geographical scope than Europe (Krzyżanowski et al. 2018, 1–14). "Crisis" narrations, perversely, denote humans flooding through lands and seas, along with peaks of illegal border crossings: problematic events that must urgently be governed, halted and prevented.

To add context to the spatial data visualisations, Migration Trail's podcast provides the audience with thorough background clarifications, situating the narrative of the two fictional characters within the structural processes of migration, and mitigating the imagined adventure described by the textual descriptions. Ultimately, the interplay of the three modes—the visual, textual and sound elements—jointly provide meaning for the whole construction, constituting, as mentioned above, a highly complementary constellation.

Finally, it is relevant to reflect on Migration Trail's relation to the general media ecosystem that shapes European public discourses on migration. In this regard, Migration Trail's dynamic interplay of fictional chats, data visualisation and cartography represent an original attempt to convey migrants' journeys in the digital realm. Nonetheless, it must be noted that migrants' narratives in the European digital media sphere reflect "an ordered space of representation and recognition" (Georgiou 2018, 55). In this domain, power relations are well cemented and often consign original, independent "spaces of witnessing" to the side-lines of the digital public sphere (Horsti 2016, 1–20). In this vein, Migration Trail certainly suffers from a lack of public visibility, evidenced by the limited popularity of the website (i.e. presented in November 2017, Migration Trail's podcast has had less than 3000 plays overall—accessed in March 2021). Such limited reach certainly dilutes the claim of potentially positively impacting the European news media coverage about migration. Once again, the diversification potential of DIY media initiatives must be carefully measured against the persistence of media hierarchies that "shape and skew coverage" (Thussu 2014, 733).

Still, when assessing the potential of such an experimental production, it might be appropriate to consider that, despite being a niche production, Migration Trail could possibly have paved the way for a new composite genre of creations being picked up in the migration-media landscape. In this spirit, the Dutch newspaper *De Correspondent* blended a long-form traditional article with screenshots of the journalist's personal WhatsApp chat with 12 migrants (Vermeulen 2020). The piece included intimate snapshots of the migrants' lives following their forced repatriation after a failed attempt to reach Europe. As in Migration Trail, a textual chat box signals an effort to humanise and personalise a traditional report on migration.

CONCLUSION

This chapter aimed at presenting an in-depth description of how Migration Trail combined data visualisation, fiction and cartography, to provide an original portrait of human mobility. More specifically, the discussion attempted to clarify how fictional elements, in combination with geospatial data visualisations, and audio contents, could challenge the processes of migrant othering present in many media productions on migration. In the first stage, this work was intended to thoroughly map Migration Trail functions, describing the website's content while understanding its dynamic interplay of multimedia elements. This allowed us to detail how the site operates, explaining how verbal, visual, textual and sound elements work together to deliver the communicative message.

The fictional chat box provides the opportunity for a prolonged emphatic involvement of the audience: prompting a protracted exposure to the story, and engaging users in a 10-day reflection on migration. In parallel, political data visualisations and the podcast give migrants' journeys a comprehensive context. They present the user with a paradoxical contrast, describing the abundance of legal connections to Europe and highlighting at the same time, how most migrants find them difficult to access. The interplay between these multiple modes of communication indicates how Migration Trail possibly transcends the label of a mere mobile-mapping exercise (Van Houtum and Bueno Lacy 2020, 196–219), effectively combining supplementary features that are commonly associated with both counter mapping and deep mapping.

A thorough analysis of the narratives embedded in the website subsequently investigated how Migration Trail confronted the ubiquitous othering of migrants. In this respect, the study highlights the attempt to re-humanise the migrants: with a clear intention to deliver the emotional, private domain of migrant lives, while rendering their expectations, fears and hopes. The narrative puts a strong emphasis on the two characters in manifest attempts to counter the symbolic duality of silent migrants: David and Sarah are neither voiceless victims, nor threatening others. They are industrious people, constantly negotiating between their original aspirations and capabilities: human beings who have set objectives and life goals and are striving to reach their destinations and achieve their purposes (de Haas 2011).

At the same time, a careful evaluation of Migration Trail's narrative intent highlights important representational shortcomings that possibly weaken its original commitment. Three levels of ethical, political and methodological concerns signal the reoccurrence of dangerous, stereotypical frames: avoidable portrayals of migrants that do not fall too far from ordinary media coverage. To illustrate, certain traits of the Migration Trail fictional narrative seem to recycle conventional ideas of migrants' deservingness: granting exclusive visibility to individuals who conform to traditional European or Western imaginaries. In parallel, the fictional narrative side-lines any expression of political statements, unconsciously naturalising the migrants' conditional existence in Europe. Thirdly, despite the innovative first-person narrative obtained through fictive SMS conversations, the account emerges as an attempt to talk about migrants rather than letting them speak. In this regard, other collaborative examples appear to have more proficiently cooperated with migrants, academics and storytellers to create truly authentic spaces of witnessing.

Despite these shortcomings, Migration Trail undoubtedly represents an original attempt to re-humanise migrants' narratives in interactive spatial data visualisations. To this end, maps were carefully conceived to represent powerful contextual elements. They help Migration Trail to challenge the Eurocentric narrative on the risks associated with non-EU immigration. Alison Killing assumes the migrants' subjective point of view and, with visual and sound signifiers, her project highlights the limited legal and financial opportunities from which migrants can benefit. Political maps recall the exclusive nature of EU border policies, a framework that illegalises human mobilities and marginalises migrants as excluded bodies.

Overall, Migration Trail emerges as a website that fruitfully narrates two compelling stories that mix political data visualisations, interactive cartography, fiction and a podcast: an exceptionally hybrid creation that could set a positive precedent for a new strand of composite narrative efforts. It could potentially provide a useful template for future efforts to narrate urgent themes that, like migration, have strong human, political and geographical components, such as climate change or the rise of global economic inequalities.

NOTES

1. "TOPOLOGICAL ATLAS" accessed 8 March 2021. http://www.topolo gicalatlas.net/.
2. "407 Camps" Mahaut Lavoine, 21 May 2019. https://mahautlavoine. com/index.php/407-camps-index/; Kaneza Schaal, 2020. http://kan ezaschaal.com/works/cartography/; "Crossing Maps". The antiAtlas of borders, 13 December 2017. https://www.antiatlas.net/crossing-maps/.
3. "Migration Trail," Alison Killing, 2018, accessed 8 March 2021. https:// www.migrationtrail.com/.

BIBLIOGRAPHY

Aarssen, Nicole. 2017. "Re-Orienting Refugee Representation?" *Stream: Interdisciplinary Journal of Communication* 9 (2): 1–14. Accessed March 8, 2021. https://journals.sfu.ca/stream/index.php/stream/article/view/229.
Adams, William C. 2015. "Conducting Semi-Structured Interviews." In *Handbook of Practical Program Evaluation*, edited by Kathryn E. Newcomer, Harry P. Hatry, and Joseph S. Wholey, ch. 19: 492–505. Hoboken, NJ, USA: John Wiley & Sons, Inc. https://doi.org/10.1002/9781119171386.
Agamben, Giorgio. 1998. *Sovereign Power and Bare Life*. Homo Sacer 1. Stanford, CA: Stanford University Press.
Allen, William L. 2020. "Mobility, Media, and Data Politics." In *The SAGE Handbook of Media and Migration*, by Kevin Smets, Koen Leurs, Myria Georgiou, Saskia Witteborn, and Radhika Gajjala, 180–191. London: Sage. https://doi.org/10.4135/9781526476982.n23.

Amores, Javier J., Carlos Arcila-Calderón, and Beatriz González-de-Garay. 2020. "The Gendered Representation of Refugees Using Visual Frames in the Main Western European Media." *Gender Issues* 37 (4). Accessed December 2020, 291–314. https://doi.org/10.1007/s12147-020-09248-1.

Berghahn, Daniela. 2020. "Immigrant Families in European Cinema." In *The SAGE Handbook of Media and Migration*, by Kevin Smets, Koen Leurs, Myria Georgiou, Saskia Witteborn, and Radhika Gajjala, 399–410. London: Sage. https://doi.org/10.4135/9781526476982.n40.

Berry, Mike, Inaki Garcia-Blanco, and Kerry Moore. 2016. "Press Coverage of the Refugee and Migrant Crisis in the EU: A Content Analysis of Five European Countries." 277. Accessed March 8, 2021. https://www.unhcr.org/56b b5a876.pdf.

Caine, Vera, M. Shaun Murphy, Andrew Estefan, D. Jean Clandinin, Pamela Steeves, and Janice Huber. 2017. "Exploring the Purposes of Fictionalization in Narrative Inquiry." *Qualitative Inquiry* 23 (3) (March): 215–221. https://doi.org/10.1177/1077800416643997.

Caquard, Sébastien, and William Cartwright. 2014. "Narrative Cartography: From Mapping Stories to the Narrative of Maps and Mapping." *The Cartographic Journal* 51 (2) (May): 101–106. https://doi.org/10.1179/000870 4114Z.000000000130.

Chion, Michel, and Claudia Gorbman. 2019. *Audio-Vision: Sound on Screen*. 2nd ed. New York: Columbia University Press.

Cho, Jeasik, and Allen Trent. 2006. "Validity in Qualitative Research Revisited." *Qualitative Research* 6 (3) (August): 319–340. https://doi.org/10.1177/1468794106065006.

Chouliaraki, Lilie. 2006. *The Spectatorship of Suffering*. London: Sage. https://doi.org/10.4135/9781446220658.

Chouliaraki, Lilie, Myria Georgiou, and Rafal Zaborowski. 2017. *The European 'Migration Crisis' and the Media A Cross-European Press Content Analysis*. London: London School of Economics.

Clandinin, D. Jean. 2013. *Engaging in Narrative Inquiry*. Developing Qualitative Inquiry, volume 9. Walnut Creek, CA: Left Coast Press, Inc.

Cosgrove, S. 2009. "WRIT101: Ethics of Representation for Creative Writers." *Pedagogy: Critical Approaches to Teaching Literature, Language, Composition, and Culture* 9 (1) (January 1): 134–141. https://doi.org/10.1215/15314200-2008-021.

Crampton, Jeremy W. 2009. "Cartography: Maps 2.0." *Progress in Human Geography* 33 (1) (February): 91–100. https://doi.org/10.1177/030913250809 4074.

Crawford, Kate, Kate Miltner, and Mary L. Gray. 2014. "Critiquing Big Data: Politics, Ethics, Epistemology." *International Journal of Communication* 8.

De Bruyn, Ben. 2020. "The Great Displacement: Reading Migration Fiction at the End of the World." *Humanities* 9 (1) (March 9): 25. https://doi.org/10.3390/h9010025.

De Haas, Hein. 2011. "The Determinants of International Migration: Conceptualising Policy, Origin and Destination Effects." March, Amsterdam Institute for Social Science Research 32. Accessed March 8, 2021. https://www.migrationinstitute.org/publications/wp-32-11.

Drüeke, Ricarda, Elisabeth Klaus, and Anita Moser. 2021. "Spaces of Identity in the Context of Media Images and Artistic Representations of Refugees and Migration in Austria." *European Journal of Cultural Studies* 24 (1) (February): 160–183. https://doi.org/10.1177/1367549419886044.

Geertz, Clifford. 1973. *The Interpretation of Cultures: Selected Essays*. Nachdr. Basic Book-s. New York: Basic Books.

Georgiou, Myria. 2018. "Does the Subaltern Speak? Migrant Voices in Digital Europe." *Popular Communication* 16 (1) (January 2): 45–57. https://doi.org/10.1080/15405702.2017.1412440.

Harley, J. B., and Paul Laxton. 2002. *The New Nature of Maps: Essays in the History of Cartography*. Johns Hopkins Paperbacks. Baltimore, MD: Johns Hopkins University Press.

Hiippala, Tuomo. 2020. *A Multimodal Perspective on Data Visualization*, edited by Martin Engebretsen and Kennedy Helen. Amsterdam Nederland: Amsterdam University Press. https://doi.org/10.5117/9789463722902.

Hiltunen, Kaisa. 2019. "Recent Documentary Films about Migration: In Search of Common Humanity." 16.

Horsti, Karina. 2016. "Visibility without Voice: Media Witnessing Irregular Migrants in BBC Online News Journalism." *African Journalism Studies* 37 (1) (January 2): 1–20. https://doi.org/10.1080/23743670.2015.1084585.

Killing, Alison. 2018. "Building Digital Stories: Architecture and Cartography Meet Documentary and Journalism." *Architectural Design* 88 (5) (September): 30–37. /https://doi.org/10.1002/ad.2339.

Krzyżanowski, Michał, Anna Triandafyllidou, and Ruth Wodak. 2018. "The Mediatization and the Politicization of the 'Refugee Crisis' in Europe." *Journal of Immigrant & Refugee Studies* 16 (1–2) (April 3): 1–14. https://doi.org/10.1080/15562948.2017.1353189.

Lo Presti, Laura. 2020 . "The Migrancies of Maps: Complicating the Critical Cartography and Migration Nexus in 'Migro-Mobility' Thinking." *Mobilities* 15 (6) (November 1): 911–929. https://doi.org/10.1080/17450101.2020.1799660.

Madörin, Anouk. "'The View from above' at Europe's Maritime Borders: Racial Securitization from Visuality to Postvisuality." *European Journal of Cultural Studies* 23 (5) (October 2020): 698–711. https://doi.org/10.1177/136754 9419869356.

Mandelbaum, Jacques, and Philippe Ridet. 2011. "L'immigré, Vedette Américaine De La Mostra De Venise." *Le Monde*. September 10, 2011. Accessed March 8, 2021. https://www.lemonde.fr/cinema/article/2011/09/10/l-immigre-vedette-americaine-de-la-mostra-de-venise_1570354_3476.html.

Martinec, R. 2005. "A System for Image-Text Relations in New (and Old) Media." *Visual Communication* 4 (3) (October 1): 337–371. https://doi.org/10.1177/1470357205055928.

McKinnon, Innisfree. 2020. "Expanding Cartographic Practices in the Social Sciences." In *The Sage Handbook of Visual Research Methods*, edited by Luc Pauwels and Dawn Mannay, 22. London: Sage.

Mekdjian, Sarah, and Anne-Laure Amilhat Szary. 2018. "Counter–Cartographies of Exile." In *This Is Not an Atlas*, 258–63. transcript Verlag. https://doi.org/10.14361/9783839445198-033.

Migreurop. 2017. *Atlas Des Migrants En Europe: Approches Critiques Des Politiques Migratoires*. Paris: Armand Colin.

Mills, Claudia. 2000. "Appropriating Others' Stories: Some Questions about the Ethics of Writing Fiction." *Journal of Social Philosophy* 31 (2) (May): 195–206. https://doi.org/10.1111/0047-2786.00041.

Mills, Claudia. 2004. "Friendship, Fiction, and Memoir: Trust and Betrayal in Writing from One's Own Life." In *The Ethics of Life Writing*, edited by J. P. Eakin, 101–120. Cornell, NY: Cornell University Press.

Monmonier, Mark S. 2018. *How to Lie with Maps*. 3rd ed. Chicago, London: The University of Chicago Press.

Nikunen, Kaarina. 2020. "Breaking the Silence: From Representations of Victims and Threat towards Spaces of Voice." In *The SAGE Handbook of Media and Migration*, by Kevin Smets, Koen Leurs, Myria Georgiou, Saskia Witteborn, and Radhika Gajjala, 411–423. London: Sage. https://doi.org/10.4135/978 1526476982.n41.

O'Halloran, Kay L., and Bradley A. Smith. 2012. "Multimodal Text Analysis." In *The Encyclopedia of Applied Linguistics*, edited by Carol Chapelle, wbeal0817. Hoboken, NJ, USA: John Wiley & Sons, Inc.. https://doi.org/10.1002/9781405198431.wbeal0817.

Papadopoulos, Dimitris, and Vassilis S. Tsianos. 2013. "After Citizenship: Autonomy of Migration, Organisational Ontology and Mobile Commons." *Citizenship Studies* 17 (2) (April): 178–196. https://doi.org/10.1080/136 21025.2013.780736.

Pauwels, Luc. 2012. "A Multimodal Framework for Analyzing Websites as Cultural Expressions." *Journal of Computer-Mediated Communication* 17 (3) (April): 247–265. https://doi.org/10.1111/j.1083-6101.2012.01572.x.

Peluso, Nancy Lee. 1995. "Whose Woods Are These? Counter-Mapping Forest Territories in Kalimantan, Indonesia." *Antipode* 27 (4) (October): 383–406. https://doi.org/10.1111/j.1467-8330.1995.tb00286.x.

Peterle, Giada. 2019. "Carto-Fiction: Narrativising Maps through Creative Writing." *Social & Cultural Geography* 20 (8) (October 13): 1070–1093. https://doi.org/10.1080/14649365.2018.1428820.

Ponzanesi, Sandra. 2016. "On the Waterfront: Truth and Fiction in Postcolonial Cinema from the South of Europe." *Interventions* 18 (2) (March 3): 217–233. https://doi.org/10.1080/1369801X.2015.1079501.

Rall, Katharina, Margaret L. Satterthwaite, Anshul Vikram Pandey, John Emerson, Jeremy Boy, Oded Nov, and Enrico Bertini. 2016. "Data Visualization for Human Rights Advocacy." *Journal of Human Rights Practice* 8 (2) (July): 171–197. https://doi.org/10.1093/jhuman/huw011.

Risam, Roopika. 2019. "Beyond the Migrant 'Problem': Visualizing Global Migration." *Television & New Media* 20 (6) (September): 566–580. https://doi.org/10.1177/1527476419857679.

Ruppert, Evelyn, Engin Isin, and Didier Bigo. 2017. "Data Politics." *Big Data & Society* 4 (2) (December): 205395171771774. https://doi.org/10.1177/2053951717717749.

Shrestha, Maheshwor. 2020. "Get Rich or Die Tryin': Perceived Earnings, Perceived Mortality Rates, and Migration Decisions of Potential Work Migrants from Nepal." *The World Bank Economic Review* 34 (1) (February 1): 1–27. https://doi.org/10.1093/wber/lhz023.

Tacchi, Jo. 2012. "Digital Engagement: Voice and Participation in Development." *Digital Anthropology*, edited by Heather A. Horst and Daniel Miller, 1st ed., 225–241. London: Routledge. https://doi.org/10.4324/9781003085201-15.

Tazzioli, Martina. 2019. "Counter-Mapping, Refugees and Asylum Borders." In *Handbook on Critical Geographies of Migration*, by Katharyne Mitchell, Reece Jones, and Jennifer Fluri, 397–409. Edward Elgar Publishing. https://doi.org/10.4337/9781786436030.00043.

Thussu, Daya. 2014. "Book Review: Lilie Chouliaraki, The Ironic Spectator: Solidarity in the Age of Post-Humanitarianism." *Media, Culture & Society* 36 (5) (July): 732–734. https://doi.org/https://doi.org/10.1177/1354066114539441.

Udah, Hyacinth, and Parlo Singh. 2019. "Identity, Othering and Belonging: Toward an Understanding of Difference and the Experiences of African Immigrants to Australia." *Social Identities* 25 (6) (November 2): 843–859. https://doi.org/10.1080/13504630.2018.1564268.

Van Heelsum, Anja. 2016. "Why Migration Will Continue: Aspirations and Capabilities of Syrians and Ethiopians with Different Educational Backgrounds." *Ethnic and Racial Studies* 39 (8) (June 20): 1301–1309. https://doi.org/10.1080/01419870.2016.1159711.

Van Houtum, Henk, and Rodrigo Bueno Lacy. 2020. "The Migration Map Trap. On the Invasion Arrows in the Cartography of Migration." *Mobilities* 15 (2) (March 3): 196–219. https://doi.org/10.1080/17450101.2019.1676031.

van Leeuwen, Theo. 2011. "Multimodality and Multimodal Research." In *The SAGE Handbook of Visual Research Methods*, by Eric Margolis and Luc Pauwels, 549–569. London: Sage. https://doi.org/10.4135/9781446268278.n28.

Vermeulen, Maite. "What Happens to Migrants Who Are Sent Back? I Spent a Year Following People to Find Out." *The Correspondent*, 9 January 2020. Accessed March 8, 2021. https://thecorrespondent.com/213/what-happens-to-migrants-who-are-sent-back-i-spent-a-year-following-12-people-to-find-out/28168874481-35612b42.

Zaborowski, Rafal, and Myria Georgiou. 2019. "Gamers versus Zombies? Visual Mediation of the Citizen/Non-Citizen Encounter in Europe's 'Refugee Crisis'." *Popular Communication* 17 (2) (April 3): 92–108. https://doi.org/10.1080/15405702.2019.1572150.

CHAPTER 5

Migration Multiple? Big Data, Knowledge Practices and the Governability of Migration

Laura Stielike

INTRODUCTION

The production of knowledge on migration is a growing field of both institutional practice and academic research. On the one hand, there is a *"migration knowledge hype"* (Braun et al. 2018, 9) among states, international organisations, and non-governmental organisations. An example of this "hype" is the recent establishment of three international knowledge hubs on migration: IOM's Global Migration Data Analysis Centre (2015), the European Commission's Knowledge Centre on Migration and Demography (2016) and the UNHCR-World Bank's Joint Data Center on Forced Displacement (2018). On the other hand, migration research itself is increasingly focusing on the production of knowledge in the field of migration governance (Casas-Cortes and Cobarrubias 2018; Nash 2018; Bartels 2018; Boswell 2009; Boswell et al. 2011) and on the critical analysis of its own processes of knowledge production (Nieswand and Drotbohm 2014; Dahinden 2016; Hatton 2018).

L. Stielike (✉)
University of Osnabrück, Osnabrück, Germany
e-mail: laura.stielike@uni-osnabrueck.de

© The Author(s) 2022   113
M. Sandberg et al. (eds.), *Research Methodologies and Ethical Challenges in Digital Migration Studies*, Approaches to Social Inequality and Difference, https://doi.org/10.1007/978-3-030-81226-3_5

This chapter will focus on the big-data-based production of knowledge on migration. Recently, migration has become an object of study for data scientists and computational social scientists employing algorithms to analyse social media data, search engine data and mobile-phone positioning data. The big-data-based production of knowledge takes place in universities and public research institutions, in data hubs of international organisations, in Internet and technology companies as well as in NGOs and private–public partnership arrangements. The urgent calls by states and international organisations for better migration data and more evidence-based migration policy in recent years, as well as a growing information and communication technology sector, have fuelled this development.

So far, this new development has been studied through the lens of data challenges in the name of development and humanitarianism (Taylor 2016), through the lens of the market-making strategies of big-data analytics firms (Taylor and Meissner 2020) as well as through the lens of a reassembling of methods by statisticians working in national statistical offices confronted with new big-data-based migration statistics (Ruppert and Scheel 2019). In this chapter, I propose to explore this development through the lens of the emergence of a new sub-discipline of migration research. Therefore, my analysis will focus on big-data-based research papers on migration produced by data scientists and computational social scientists based at universities and public research institutions. Following Annemarie Mol (2002) and Stephan Scheel et al. (2019), I will explore how migration is enacted through data practices, and precisely how migration and migrants are enacted through big-data-based research papers. I will argue that migration and migrants are enacted in multiple ways and that this multiplicity—or inconsistency—is held together by reference to three mainstream migration narratives—demography, integration and humanitarianism—which all frame migration as something that needs to be governed and that can be governed better through better data. With this study, I would like to contribute to the larger debate on the role of scientific knowledge production for migration policy and governance (Boswell 2009; Boswell et al. 2011; Geddes 2015).

In the first section of this chapter, I will frame the emerging transnational network of actors involved in the big-data-based production of knowledge on migration as an apparatus that emerged in response to two discourses of urgency related to the crisis of migration governance and the UN Sustainable Development Goals. The second section reflects on a praxiographic approach to studying the production of knowledge on migration. In the third section, I present the findings of an analysis of 17 big-data-based research papers and carve out the multiple enactments of migration and how they are held together by reference to three mainstream migration narratives.

The Big Data and Migration Apparatus

In June 2018, the Knowledge Centre on Migration and Demography of the European Commission and the Global Migration Data Analysis Centre of the International Organization for Migration launched the Big Data for Migration Alliance. The aim of this alliance is to "advance discussions on how to harness the potential of big data sources for the analysis of migration and its relevance for policymaking" (Knowledge Centre on Migration and Demography and Global Migration Data Analysis Centre 2018). In December 2018, 164 states signed the Global Compact for Safe, Orderly and Regular Migration. The compact's first objective is to "collect and utilize accurate and disaggregated data as a basis for evidence-based policies" (United Nations 2018, 6). Among the proposed actions is the use of big data for the governance of international migration. What is so fascinating about big data for migration policymakers? Why would they like to make use of social media posts, web search histories and mobile-phone positioning data? Employing the term *evidence-based policy*, the official explanation presented is that the more accurate knowledge policymakers have about migration, the better they can develop policies and tools to manage it (Geddes 2015; Stielike 2017, 129ff.). In this respect, it seems promising to access big data that is virtually real-time or can be updated frequently, that covers geographic areas with no or limited official migration statistics and that has much larger sample sizes and more flexible definitions of migration than traditional surveys (Rango and Vespe 2018, 6). Of course, the use of big data for migration governance must be seen as part of a larger trend to employ algorithms for political decision-making (see e.g. Yeung and Lodge 2019).

I frame the growing interest for big-data-based migration research and governance among European migration policymakers as an effect of a newly established big data and migration apparatus. Michel Foucault describes an apparatus as "a thoroughly heterogeneous ensemble consisting of discourses, institutions, architectural forms, regulatory decisions, laws, administrative measures, scientific statements, philosophical, moral and philanthropic propositions – in short, the said as much as the unsaid" (Foucault 1980, 194). Following Foucault, I understand an apparatus as a network ("réseau") consisting of different elements such as discourses, institutions and modes of subjectivation. According to Foucault, an apparatus is established in response to a discourse of urgency and interferes in power relations that stabilise and destabilise certain types of knowledge (Foucault 1980, 194–197; Raffnsøe 2008; Agamben 2008). The big data and migration apparatus can be described as an emerging transnational network of international organisations' data hubs, data researchers at universities, internet and technology companies and non-profit organisations involved in the big-data-based production of knowledge on migration.

The big data and migration apparatus has evolved since 2015 in response to two discourses of urgency. The adoption of the Sustainable Development Goals in 2015 established a discourse of urgency related to the improvement of data on migration. Altogether, 10 of the 17 Sustainable Developments Goals contain targets and indicators of relevance to migration or mobility. So far, measuring the progress towards achieving the migration-related targets has not been possible as there is insufficient data on migration. Also, the Agenda's core principle to "leave no one behind", including migrants, requires data disaggregation by migratory status, creating significant migration data needs. Therefore, the UN Statistics Division issued an urgent call for the improvement of migration data and for innovative means of data collection, namely big data (UN Statistics Division 2017; United Nations Expert Group Meeting 2017).

The second discourse of urgency consists of four problematisations in the migration crisis discourse that evolved in the summer of 2015. The dynamics of autonomous movement of migrants across European borders questioned the states' ability to control migration; the EU's unwillingness to finance an effective search and rescue operation in the Mediterranean Sea questioned the humanitarian principles of the European Union; the chaotic situations with regard to registration and accommodation of migrants questioned government agencies' ability to properly integrate migrants; and the populist debates on migration in the aftermath of 2015 challenged the objectivity of information on migration. The use of big data for the study and governance of international migration promises to resolve the crisis of European migration governance as it responds to all four problematisations: control, humanitarianism, integration and objectivity. Policymakers believe that big data can be used to strengthen migration control through the monitoring, forecasting and now-casting of migration dynamics, to improve humanitarian action related to migration, to enhance integration policies and to deliver objective information on migration.[1]

ENACTING MIGRATION: KNOWLEDGE PRACTICES AND MIGRATION MULTIPLE

A common question asked in big-data-based migration research is whether the data used represents reality. To verify this, data scientists and computational social scientists often compare their results to conventional statistics. They conceive conventional statistics as a "gold standard" that represents the "ground truth" or reality as closely as possible. Following the thoughts of Annemarie Mol, we can argue that there is not only one reality. In Mol's praxiographic perspective, reality is multiple and done in practice. Instead of asking the epistemological question of whether representations of reality are accurate, she studies the ways in which "objects come into being – and disappear – with the practices in which they are manipulated" (Mol 2002, 5). She calls this coming-into-being *enactment*: "It is possible to say that in practices objects are *enacted*" (Mol 2002, 33, emphasis in original). This shift from asking "how to find the truth?" to "how are objects handled in practice?" pushes the philosophy of science to develop "an *ethnographic* interest in knowledge practices" (Mol 2002, 5, emphasis in original).

Mol argues that "reality multiplies" as "objects tend to differ from one practice to another" (Mol 2002, 5). In her book *The Body Multiple*, she shows how in a Dutch hospital, the disease atherosclerosis is enacted in various diagnostic and therapeutical practices and thereby brings into existence several versions of atherosclerosis. In other words and related to the social sciences, "different research practices might be *making multiple worlds*" (Law and Urry 2004, 397, emphasis in original). John Law and John Urry speak of the "*performativity* of method", meaning that a specific research method "helps to produce the realities that it describes" (Law and Urry 2004, 397, emphasis in original).

However, if method is interactively performative, and helps to make realities, then the differences between research findings produced by different methods or in different research traditions have an alternative significance. No longer different *perspectives* on a single reality, they become instead the enactment of different *realities* (Law and Urry 2004, 397, emphasis in original).

However, in her book, Mol makes a "double move" (Mol 2002, 82). She not only studies "the multiplication of a single disease" but also "the coordination of this multitude into singularity" (Mol 2002, 82). In other words, she identifies how the different versions of atherosclerosis enacted in the hospital "hang together" (Mol 2002, 84) and identifies four "forms of coordination" (Mol 2002, 55) or "recurrent patterns of coexistence between different enactments" (Mol 2002, 181) of the disease: addition, translation, distribution and inclusion. Addition means to add different objects together and thereby turn them into one either by establishing a hierarchy or through cumulative arguments (Mol 2002, 55–72), translation means to make the results of distinct practices comparable (Mol 2002, 72–85), distribution means to keep incoherent objects separated between different sites in order to prevent a clash between them (Mol 2002, 87–117) and inclusion means that some objects mutually include and constitute each other (Mol 2002, 120–142). By focusing on these modes of coordination, Mol stresses that the singularity of an object—such as a disease or migration—is "an accomplishment" and "the result of the work of coordination" (Mol 2002, 119).

In their special issue "Enacting migration through data practices", Stephan Scheel, Evelyn Ruppert and Funda Ustek-Spilda draw on Annemarie Mol's—and John Law's (2004, 2008, 2012)—work to study the enactment of migration through data practices. More precisely, they argue that the "enactment agenda" should be put to use not only at conventional sites of knowledge production such as laboratories and hospitals but also in "politically highly contested contexts" such as migration governance (Scheel et al. 2019, 583). They understand practices not as mere techniques or technical operations but as "*activities* performed by humans in relation to materials, technologies and shared understandings" that "occur within specific fields" (Scheel et al. 2019, 583, emphasis in original). Examples of data practices are "judgements and tacit knowledge of practitioners", "rules, standards and struggles within a community of practice" as well as "the affordances and constraints of technologies" (Scheel et al. 2019, 583). Finally, they ask "how – and through what kind of data practices – migration-related realities are enacted as objects of government" (Scheel et al. 2019, 582). Consequently, most contributions to their special issue focus on sites and actors more or less directly involved in the governance of migration, such as administrative offices, refugee camps, border patrols, national statistical offices and international organisations (Plájás et al. 2019; Pollozek and Passoth 2019; van Reekum 2019; Schultz 2019; Scheel and Ustek-Spilda 2019). By comparison, this chapter chooses a more conventional site of knowledge production by focusing on big-data-based research produced by data scientists and computational social scientists at universities and public research institutions. However, I will show that even though the sites and actors are not directly involved in governmental practices, the knowledge they produce on migration is closely linked to the field of migration governance.

While building mainly on ethnographic techniques of observation and writing, Mol points out that "[a]nother quite different but equally interesting resource for praxiography is found in the *material and methods* sections of scientific articles" (Mol 2002, 158, emphasis in original). This piece of writing is part of a larger research project on the knowledge practices involved in the big-data-based production of knowledge on migration, which employs a multi-sited ethnography including observations, interviews and document analysis. However, for this chapter, I follow Mol's suggestion and draw on big-data-based research papers on migration, paying particular attention to their material and methods sections. I focus on how multiple versions of migration are enacted in these papers and how—at the same time—migration "hangs together" as a singular object. In this chapter, I will not engage in the debate on the extent to which big-data-based research (on migration) is representative, biased, legitimate, ethical or trustworthy (see e.g. Taylor and Dencik 2020; Ho 2020).

Enactments of Migration in Big-Data-Based Research Papers

The following analysis is based on 17 big-data-based research papers related to "migration" or "mobility" published between 2011 and 2020. I chose the papers for their wide range of big-data sources (call detail records, geo-coded e-mail logins, LinkedIn profiles, geo-located Twitter tweets, geo-coded Skype login data, geo-coded Google+ data, Facebook advertising platform data, Facebook interests, Facebook Network data, Google Trends Index) and for their great variety in addressing "migration" or "mobility" (international migration, international mobility, high-skilled migration, daily travelling, transnationalism, assimilation, segregation, relocation between three countries, forced mobility after natural disasters, forced migration due to economic and political crises). All the selected papers present original research. Most of them were co-authored by researchers who have been among the first to use big data to study migration, have become central figures in the field and have offered advice to government agencies or international organisations. I have not included papers that have been produced for international organisations (e.g. Hughes et al. 2016; UN Global Pulse and UNHCR Innovation Service 2017; Spyratos et al. 2018) or that provide an overview of the field but no original research (e.g. Sîrbu et al. 2020). Although the selection cannot be considered representative in a statistical sense, in my view, the selected papers give good insight into the field.

At the time of publication of the selected papers, 32 authors worked at universities or research institutes in Europe, North America, Asia, the Middle East and South America. Five authors were employed at technology companies (Microsoft, Yahoo, LinkedIn, Positium) and ten authors worked for international, supranational or non-profit organisations (OECD, the European Commission's Knowledge Centre on Migration and Demography, IOM's Global Migration Data Analysis Centre, UNICEF, iMMAP Colombia, Global Protection Cluster Switzerland). On the basis of their given names, it can be estimated that five authors are female and 40 are male.

The large majority of research papers consider the lack of timely, reliable, comparable and disaggregated data that build on consistent definitions of migration and have a wide geographical coverage as the central problem of migration research. In response to this identified problem, most papers argue that big data provide a solution. Big data sources promise timeliness, consistency, disaggregation, higher spatial resolutions, coverage of developing countries and hard-to-reach populations and the capture of forms of migration and mobility that are not represented in official statistics. The new data sources not only promise to complement traditional statistics but also to deliver data that can be used for predictive purposes such as now-casting and forecasting: "The main goals of our work are to complement existing migration statistics, and to develop methods for harnessing publicly available online data in order to improve forecasts and our understanding of populations of migrants" (Zagheni et al. 2014, 1f.).

Multiple Enactments of Migration: Data-Driven Definitions

Inconsistent definitions of migration play an important part in the recurring "lack of data" problematisation summarised above. While it is seen as a problem of official migration statistics that "different countries collect data for different purposes and thus use different definitions of migration", the possibility to "use the same definition of migration consistently" (State et al. 2013, 1) is considered a major advantage of working with big data. Thus, every research paper provides its own definition of migration or migrant based on the specific data source used for the study. This practice produces a multiplicity of definitions that are highly data driven. In the following, I focus on two definitions of migration and three definitions of migrant.

In their paper on the use of e-mail data for estimating international migration rates, Emilio Zagheni and Ingmar Weber "define migration as a change of usual residence between the period from 09-2009 to 06-2010, and the period from 07-2010 to 06-2011" (Zagheni and Weber 2012, 3). This definition builds on the data they use for their study: A large sample of Yahoo e-mail messages sent between September 2009 and June 2011. The authors know the self-reported date of birth and gender of the e-mail account holders as well as the dates the messages were sent. Based on the users' IP addresses, they can also estimate the country from where the messages were sent. Simply put, in the view of the authors, a "change of usual residence" has occurred when in the first time period, a user has sent most e-mails from country A and in the second time period from country B.

Zagheni and Weber's definition of migration resembles the definition proposed by the International Organization for Migration (IOM) inasmuch as it also refers to a "place of usual residence".[2] In contrast to IOM's definition, movement itself does not play a role in Zagheni and Weber's definition. They focus on a change of location that has occurred between two time periods but not on the process of mobility or movement itself. Unlike often-used definitions by national statistical offices.[3] Zagheni and Weber's definition is not linked to notions such as birthplace or nationality. The person who changes their usual residence between the two time periods could be a national of one of the two countries, of both of them or of a third country. Their definition of migration also does not relate to motives or determinants of migration. Implicitly, to them, migration is international migration as mentioned in the title of their paper.

Bogdan State et al. draw on data from the online platform LinkedIn to study the migration of professionals to the US. Instead of defining migration, they define a "migration event":

> We define a migration event by querying the location of each individual at the beginning of every calendar year. If the individual's estimated place of residence is in a different country, compared to the beginning of the previous year, we assume that a migration event has occurred during the past calendar year. (State et al. 2014, 540)

More precisely, they "measured migrations by examining country-level locations associated with positions held by individuals across their careers, as listed in their LinkedIn profiles" (State et al. 2014, 540). Those migrations had to last for at least one calendar year and must have taken place between 1990 and 2012. In short, a migration event is defined as a change of employment listed in a user's LinkedIn profile that is related to a changed country of residence and lasts for at least one year. Again, from this view, migration is not defined by nationality or birthplace and it is implicitly understood as international migration. Also, this definition does not focus on movement but on migration as the result of an already completed process of mobility. The criterion of length of stay of at least one year coincides with the definition of migration proposed by the European Migration Network.[4]

Interestingly, most big-data-based research papers do not define migration but migrant. Those definitions are more diverse than the definitions of migration but are also heavily data driven. Three main types of definitions can be distinguished: (1) a specific amount of time spent in at least two countries, (2) self-reported multiple places of residence, (3) self-reported and inferred residence in a country different from the original country of residence.[5]

The first type of definition considers a migrant a person who spends a specific amount of time in one country and another specific amount of time in another country. In their paper on studying international mobility through IP-address-based geo-located logins into Yahoo accounts, Bogdan State et al. define a migrant as "an individual who spends at least 90 days in exactly two countries during the observed timespan of one year" (State et al. 2013, 3). Length of stay and the stay in at least two countries are also the main aspects of the definition in a paper on the use of IP-address-based geo-located logins into Skype accounts to explain international migration. Riivo Kikas et al. define Skype users as migrants "if they have been in one country for at least five consecutive months and in another country for at least five consecutive months. Setting these time limits prevents counting longer holidays or business visits as migration events" (Kikas et al. 2015, 18). Similarly, in their study on international and internal migration patterns inferred from geo-located Twitter tweets, Zagheni et al. define migrants as "those users that are identified as people who moved to a different country for at least one of the 4-month periods that we considered" (Zagheni et al. 2014, 4).

In their often-cited definition for collecting data on migration, the United Nations Department of Economic and Social Affairs (UN DESA) defines an *"international migrant"* as *"any person who changes his or her country of usual residence"* (UN DESA 1998, 9).[6] UN DESA also provides definitions for "long-term migrant"[7] and "short-term migrant".[8] The big-data-driven migrant definitions cited above indirectly relate to the definition of short-term migrant as the minimal length of stay is set at 90 days (approx. 3 months). However, except for the first study, they do not set a maximal length of stay, thus examining both what UN DESA defines as short-term and long-term migration. Unlike the European Migration Network, which defines a migrant as "a person who is outside the territory of the State of which they are nationals or citizens and who has resided in a foreign country for more than one year irrespective of the causes, voluntary or involuntary, and the means, regular or irregular, used to migrate" (European Migration Network 2021), this first type of big-data-driven definition does not refer to categories such as nationality, citizenship, motivations or means of migration.

Interestingly, Rein Ahas et al. also base their definition of "transnationals" on length of stay and stay in at least two countries. However, in their paper on tracking transnationalism with mobile telephone data, they add a third parameter: the number of trips. Drawing on domestic and roaming call detail record (CDR) data of the two largest mobile communications operators in Estonia for the year 2015, they consider people as transnationals "if they spend more than 25% of their time (at least 92 days), but not more than 75% of their time (up to 273 days) in a foreign country" and "if they have taken at least five trips to a foreign country, but not more than 52 trips (once a week)" (Ahas et al. 2017, 8). In summary, this first type of definition of migrants (and transnationals) resembles the above-presented definitions of migration by its reference to length of stay and stay in at least two countries as well as in its non-reference to categories such as country of birth, country of origin, nationality or citizenship.

The second type of definition relates to multiple places of residence reported by users. In their paper, Johnnatan Messias et al. draw on data from Google+ profiles to study migration clusters—the relocation of a person between three countries. Users of Google+ accounts can list in their profiles all the places in the world where they have lived. These "Places where I lived" are automatically geo-coded by Google+. Unlike in the first type of definition, where migration is inferred from a change in geo-coded logins of users, here the definition of migrant builds on multiple former places of residence self-reported by the users: "As our study is about international migration, we only considered the subset of users who have lived ('places lived') in at least two distinct countries. We refer to this group of users as *migrants*" (Messias et al. 2016, 423, emphasis in original). This means that people are considered as migrants if they have ever lived in more than one country. Even if they have returned to their country of origin after a short period of time abroad, they are still considered to be migrants. Thus, everyone who has studied abroad for a semester or worked for a year in a foreign country and returned is considered a migrant for the rest of their lives. This understanding of migrant stands in stark contrast to the use of the term in public discourse or national statistical offices. Here "migrants" are only those who have arrived from abroad—and sometimes even their children are marked by the German statistical category "migration background" (Will 2019)— but not those who have returned from abroad. This second type of definition resembles the first type in lacking any reference to categories such as country of origin or nationality, but it differs from the first type in also lacking any reference to the length of stay.

The third type of migrant definition is based on a mixture of user-reported and inferred information on the users' residence outside their "original country of residence". Drawing on data from Facebook's advertising platform, Antoine Dubois et al. explore "migrant assimilation through Facebook interests". Using Facebook's Marketing API, companies and researchers can obtain estimates of the number of users "who belong to a certain demographic group and show certain *interests*" (Dubois et al. 2018, 53, emphasis in original). As Facebook does not provide the category "migrants", the authors use Facebook's category "expats" instead: "We use the Facebook advertising platform terminology, which does not refer to *migrants* but to expats, though we use migrant and expat interchangeably" (Dubois et al. 2018, 53, footnote 7). Facebook defines expats as "people whose original country of residence is different from the current country".[9] Facebook does not provide information on how users are categorised as "expats". However, Zagheni et al. infer from research produced by Facebook staff that the "current city" and "hometown" provided by users in their Facebook profiles as well as the structure of the users' network of Facebook friends must be "among the key components of the estimation process" (Zagheni et al. 2017, 724).[10] In their paper on the quantification of human mobility patterns using Facebook Network data, Spyratos et al. indicate that the Facebook-based definition of migrants does not refer "to a user's citizenship, country of birth, or legal status" (Spyratos et al. 2019, 5). However, unlike the first and second type, this third type of definition refers to an "original country of residence" and a "hometown", thus using categories that are close to country of origin or country of birth. As in the second type, there is no reference to the length of stay.

As shown above, big-data-based research papers enact migration and migrants in multiple ways. The two presented definitions of migration are not so diverse, as both build on two main criteria: change of usual residence to another country and length of stay. However, they differ in the defined length of stay. Both definitions do not relate to categories such as country of origin or nationality. The definitions of migrants presented in the research papers are manifold. While the first type of definition builds on length of stay and stay in at least two countries, similar to the criteria for the definitions of migration, the second and third types of definition do not refer to length of stay. Instead, according to the second type of definition, people are considered to be migrants if they have ever lived in more than one country. Only the third type of definition builds on the idea of a given country of origin or birth that differs from the current place of residence.

From the perspective of a "reflexive turn" (Nieswand and Drotbohm 2014; Dahinden 2016; Amelina 2021) in migration studies, the big-data-driven enactments of migration and migrant hold both potentials and risks. On the one hand, they invite reflection upon the strong associations between migration, nationality and origin in "conventional" migration research and help to rethink migration beyond these categories—as (a result of) movement in space. On the other hand, these big-data-driven enactments reproduce methodological nationalism, as the nation-state—here usually called "country"—is still (implicitly) used as the key reference point to define migration and migrant. Also, the second and third type of migrant definition discussed above hold the risk that categories such as "Places where I lived" or "expats" defined by private companies and attributed by their algorithms greatly influence migration researchers' understanding of migration.

Work of Coordination: Enacting Migration as a Singular Object by Reference to Migration Narratives

How is it possible that the presented research papers treat migration and migrants as singular objects given the diverse ways in which they define them? How is the singularity of migration and migrants as an object achieved within and across the research papers? Following Annemarie Mol, I highlight the "work of coordination" that is undertaken in the research papers and argue that migration and migrants "hang together" by reference to three mainstream migration narratives—demography, integration and humanitarianism—which all frame migration as something that needs to be governed and that can be governed better through better data.

Eight out of the 17 research papers frame migration as a demographic phenomenon (Zagheni and Weber 2012; State et al. 2013, 2014; Zagheni et al. 2014; Kikas et al. 2015; Messias et al. 2016; Zagheni et al. 2017; Dubois et al. 2018). In this view, migration is understood as a factor that changes the size and composition of a population and that can be influenced to a certain extent through political interventions. In this vein, Zagheni and Weber see international migration as an "important driver of demographic growth in many countries" (Zagheni and Weber 2012, 1), State et al. consider high-skilled migration as an "important demographic phenomenon with relevant consequences, for instance in terms of human capital formation, a central issue in the study of economic development" (State et al. 2014, 537), and Dubois et al. perceive immigration as a "stopgap measure to address population aging, which would otherwise strain the economy and public finances" (Dubois et al. 2018, 51). Moreover, some authors see their papers as a direct contribution to demographic research, for example, Zagheni et al. when they write "[I]n this article, we contribute to the development of tools and methods that leverage new data sources for demographic research" (Zagheni et al., 2017, 721; see also Zagheni et al. 2014, 1; Messias et al. 2016, 427).

Three research papers relate to a second migration narrative that is built around the assumed need to integrate migrants into receiving societies (Dubois et al. 2018; Stewart et al. 2019; Marquez et al. 2019). Interestingly, two of the papers do not focus on integration into the labour market or the education system but on "cultural assimilation" in terms of "interests" expressed on Facebook (Dubois et al. 2018, 52; Stewart et al. 2019, 3258). Thus, integration is imagined as a unidirectional process of adaptation by migrants and their descendants to the population of the receiving country. In their study on the segregation between Syrian refugees and the native population in Turkey, which is based on call detail records and Twitter, Neal Marquez et al. show a significant positive relationship between positive sentiments towards refugees in Turkey expressed on Twitter and the probability of refugees contacting non-refugees via mobile phone (Marquez et al. 2019, 276). This implies that the receiving society plays its part in the integration process. However, the main responsibility for integration seems to lie with the migrants as it is their calls to non-refugees that are counted as a proxy for integrative behaviour—and not the calls of non-refugees to refugees. Thus, in all three papers, integration is primarily imagined as a one-way street.

The third migration narrative concerns humanitarian assistance to people fleeing from natural disasters, or economic, political or medical crises (Bengtsson et al. 2011; Lu et al. 2012; Blanford et al. 2015; Böhme et al. 2020; Palotti et al. 2020). The central assumption is that better data on the number, spatial distribution and routes of fleeing populations allows for improved humanitarian assistance. In their paper on the spatial distribution and socio-economic status of Venezuelan "refugees and migrants" in different receiving countries, which is based on data from the Facebook advertising platform, Joao Palotti et al. write: "Estimating the absolute number and the spatial distribution of Venezuelan refugees and migrants are (sic) a top priority in order to quantify the magnitude of the crisis and to plan an appropriate humanitarian response" (Palotti et al. 2020, 6). Linus Bengtsson et al., whose study tracks the movement of people after the earthquake in Haiti in 2010 via call detail records, also argue that the provision of close to real-time data "on postdisaster population distributions can potentially enable improved distribution of water, food, shelter, and sanitation" (Bengtsson et al. 2011, 7; see also Lu et al. 2012, 11580). Additionally, referring to the cholera outbreak in Haiti a few months after the earthquake, they show that call detail record data can also be used to "potentially inform outbreak preparedness and response for infectious diseases" (Bengtsson et al. 2011, 7). Justine Blanford et al. also point out the potential of geo-located Twitter tweets for understanding epidemic dynamics and enhancing disease surveillance (Blanford et al. 2015, 11). Even Marcus Böhme et al. see in their study on the prediction of international migration via online search keywords an approach that could "be used for policy applications in the case of humanitarian crises in order to deliver real-time monitoring of migration intentions ahead of their realization to organize humanitarian responses" (Böhme et al. 2020, 19).

All three migration narratives frame migration as something that needs to be governed and that can be governed better through better data. From a demographic perspective, better data on migration allow for better demographic forecasts and more appropriate population-related policies; from an integrationist perspective, better data on migration allow for better integration policies and from a humanitarian perspective, better data on migration allow for better planning and implementation of humanitarian assistance. Several authors consider their own research as "input for policy-making" and envisage a "systematic use of non-traditional data for policy support and migration governance" (Spyratos et al. 2019, 19).

Finally, I argue in the sense of the "performativity of methods" that the multiple big-data-driven definitions of migration and migrants discussed above enact realities beyond a governmental discourse on migration—for example, a gradual shift from state/nation/origin-centred migration thinking to mobility-centred migration thinking. For (self-)reflexive migration studies, it might be worth exploring these enactments more closely to discover alternative ways of rethinking migration. However, contrary to a pure "performativity of methods" standpoint, I also argue that the authors of the big-data-based research papers—perhaps to gain credibility and prestige as a new sub-field of migration studies—aim at contributing to the well-established research fields of demography, integration or humanitarianism and thereby inscribe into migration narratives that stand in stark contrast to these alternative enactments of migration. In this process, the research papers—some more implicitly and others very explicitly—adopt the assumption that migration needs to be managed or governed and that this can be improved through better data. Finally, this common assumption is what makes the multiple big-data-driven versions of migration and migrants "hang together".

Conclusion

In a pre-recorded online panel discussion titled "Data for what? A conversation with policymakers and practitioners on the use of evidence and data on forced displacement" that was part of the virtual United Nations World Data Forum 2020, Björn Gillsäter, head of the recently founded World Bank—UNHCR Joint Data Center on Forced Displacement, said in his introduction: "I think one of the things that unites those of us who are watching this video is that we believe in what gets measured gets done or at least what gets measured gets managed".[11] Just like the big-data-based research papers analysed in this chapter, this statement builds on the assumption that migration is an object of government, and that it needs to be managed. To make this assumption more explicit, we could reformulate it as: What needs to be governed gets measured to be governed better.

Drawing on 17 big-data-based research papers, I showed in this chapter that the emerging sub-discipline of big-data-based migration research enacts migration and migrants in multiple ways. While some papers focus on change of residence and length of stay, others define migrants by self-reported multiple former places of residence or by a mixture of self-reported and inferred residence in a country different from a supposed original country of residence. Interestingly, nationality, citizenship or country of birth hardly play a role in these enactments, which is what makes them—to a certain extent—differ from realities enacted by social science migration researchers or by actors involved in migration governance. However, following Annemarie Mol, I have argued that this multiplicity of migration is held together by reference to three migration narratives—demography, integration and humanitarianism—which all frame migration as something that needs to be governed and that can be governed better through better data. As the research papers aim at contributing to these research fields, they inscribe themselves into these migration narratives and thereby adopt the assumption of migration as an object of government. The will or necessity of data scientists and computational social scientists to relate to dominant migration narratives—perhaps to gain credibility and prestige as a new sub-field of migration studies—seems to be stronger than the "performativity of methods" that creates new migration realities. However, I would argue that—from a (self-) reflexive migration studies perspective—exactly these big-data-driven alternative enactments of migration might be worth exploring in more detail as they promise to offer new ways of rethinking migration beyond governmental discourse. Finally, we could ask in the sense of "ontological politics" (Mol 1999, 2002; Law and Urry 2004, 396f.): If methods help to make realities, which migration realities might big-data-based migration research want to enact in the future?

NOTES

1. See, for example, the initiative "Migration 4.0" organised during Germany's presidency of the Council of the European Union which covered control (e.g. forecasting tools, facial and voice recognition), humanitarianism (virtual psycho-social counselling), integration (new digital communication channels with migrants) and objectivity (better evidence through data collaboratives; study of public attitudes on migration) (German Federal Ministry of the Interior, Building and Community 2020).
2. The International Organization for Migration defines migration as "[t]he movement of persons away from their place of usual residence, either across an international border or within a State" (International Organization for Migration, n.d.).
3. For the UK's Office of National Statistics see Anderson and Blinder (2019).
4. The European Migration Network is a network of "migration and asylum experts" initiated by the European Commission's Directorate-General Migration and Home Affairs. In the network's glossary, migration in "the global context" is defined as the "movement of a person either across an international border (international migration), or within a state (internal migration) for more than one year irrespective of the causes, voluntary or involuntary, and the means, regular or irregular, used to migrate" (European Migration Network, n.d.).
5. It would need further research based on a larger selection of research papers to investigate how migration/migrant definitions and the choice of big data sources have changed over the last ten years and how this might have been related to changes in migration narratives.
6. The IOM defines a migrant as "a person who moves away from his or her place of usual residence, whether within a country or across an international border, temporarily or permanently, and for a variety of reasons" (International Organization for Migration, n.d.).
7. "A person who moves to a country other than that of his or her usual residence for a period of at least a year (12 months), so that the country of destination effectively becomes his or her new country of usual residence" (UN DESA 1998, 10).

8. "A person who moves to a country other than that of his or her usual residence for a period of at least 3 months but less than a year (12 months) except in cases where the movement to that country is for purposes of recreation, holiday, visits to friends and relatives, business, medical treatment or religious pilgrimage" (UN DESA 1998, 10).

9. Facebook Adverts Manager's documentation cited by Zagheni et al. (2017, 723).

10. In October 2018, Facebook's advertising platform changed its classification from "expats of country X" to "lived in country X" whereby users who have "lived in country X" are defined as "people who used to live in country X and now live abroad". The classification was changed back to "expats" in late 2018, while its definition remained the same (Spyratos et al. 2019, 4; Palotti et al. 2020, 10f.).

11. This video was available at the United Nations World Data Forum 2020 which took place from 19 to 21 October 2020 as a virtual event due to the corona pandemic. The quote can be found at time code 0:35 (*Data for What? A Conversation with Policymakers and Practitioners on the Use of Evidence and Data on Forced Displacement*, n.d.).

Bibliography

Agamben, Giorgio. 2008. *Was ist ein Dispositiv?* Zürich and Berlin: Diaphanes.

Ahas, Rein, Siiri Silm, and Margus Tiru. 2017. "Tracking Trans-Nationalism with Mobile Telephone Data." In *Estonian Human Development Report 2016/2017. Estonia at the Age of Migration*, edited by Tiit Tammaru, Raul Eamets, and Kristina Kallas. Foundation Estonian Cooperation Assembly. https://2017.inimareng.ee/en/open-to-the-world/tracking-transnationalism-with-mobile-telephone-data/.

Amelina, Anna. 2021. "After the Reflexive Turn in Migration Studies: Towards the Doing Migration Approach." *Population, Space and Place* 27 (1). https://doi.org/10.1002/psp.2368.

Anderson, Bridget, and Scott Blinder. 2019. "Who Counts as a Migrant? Definitions and Their Consequences." *Migration Observatory Briefing, COMPAS, University of Oxford* (blog). July 2019. https://migrationobservatory.ox.ac.uk/resources/briefings/who-counts-as-a-migrant-definitions-and-their-con sequences/.

Bartels, Inken. 2018. "Practices and Power of Knowledge Dissemination. International Organizations in the Externalization of Migration Management in Morocco and Tunisia." *Journal for Critical Migration and Border Regime Studies* 4 (1): 47–66.

Bengtsson, Linus, Xin Lu, Anna Thorson, Richard Garfield, and Johan von Schreeb. 2011. "Improved Response to Disasters and Outbreaks by Tracking Population Movements with Mobile Phone Network Data: A Post-Earthquake Geospatial Study in Haiti." Edited by Peter W. Gething. *PLoS Medicine* 8 (8): e1001083. https://doi.org/10.1371/journal.pmed.1001083.

Blanford, Justine I., Zhuojie Huang, Alexander Savelyev, and Alan M. MacEachren. 2015. "Geo-Located Tweets. Enhancing Mobility Maps and Capturing Cross-Border Movement." Edited by Renaud Lambiotte. *PLOS ONE* 10 (6): e0129202. https://doi.org/10.1371/journal.pone.0129202.

Böhme, Marcus H., André Gröger, and Tobias Stöhr. 2020. "Searching for a Better Life: Predicting International Migration with Online Search Keywords." *Journal of Development Economics* 142 (January): 102347. https://doi.org/10.1016/j.jdeveco.2019.04.002.

Boswell, Christina. 2009. *The Political Uses of Expert Knowledge: Immigration Policy and Social Research*. Cambridge and New York: Cambridge University Press.

Boswell, Christina, Andrew Geddes, and Peter Scholten. 2011. "The Role of Narratives in Migration Policy-Making: A Research Framework." *The British Journal of Politics and International Relations* 13 (1): 1–11. https://doi.org/10.1111/j.1467-856X.2010.00435.x.

Braun, Katherine, Fabian Georgi, Robert Matthies, Simona Pagano, Mathias Rodatz, and Maria Schwertl. 2018. "Umkämpfte Wissensproduktionen der Migration: Editorial." *Movements. Journal for Critical Migration and Border Regime Studies* 4 (1): 9–27.

Casas-Cortes, Maribel, and Sebastian Cobarrubias. 2018. "It Is Obvious from the Map! Disobeying the Production of Illegality beyond Borderlines." *Journal for Critical Migration and Border Regime Studies* 4 (1): 29–44.

Dahinden, Janine. 2016. "A Plea for the 'De-Migranticization' of Research on Migration and Integration." *Ethnic and Racial Studies* 39 (13): 2207–2225. https://doi.org/10.1080/01419870.2015.1124129.

Data for What? A Conversation with Policymakers and Practitioners on the Use of Evidence and Data on Forced Displacement. n.d. Accessed November 27, 2020. https://www.jointdatacenter.org/data-for-what-a-conversation-with-policymakers-and-practitioners-on-the-use-of-evidence-and-data-on-forced-displacement/.

Dubois, Antoine, Emilio Zagheni, Kiran Garimella, and Ingmar Weber. 2018. "Studying Migrant Assimilation Through Facebook Interests." In *Social Informatics*, edited by Steffen Staab, Olessia Koltsova, and Dmitry I. Ignatov, 11186: 51–60. Cham: Springer International Publishing. https://doi.org/10.1007/978-3-030-01159-8_5.

European Migration Network. 2021. "Migrant." Glossary. Accessed March 15, 2021. https://ec.europa.eu/home-affairs/what-we-do/networks/european_migration_network/glossary_search/migration_en.

European Migration Network. n.d. "Migration." Glossary. Accessed March 15, 2021. https://ec.europa.eu/home-affairs/what-we-do/networks/european_migration_network/glossary_search/migration_en.

Foucault, Michel. 1980. "The Confession of the Flesh [1977]." In *Power/Knowledge: Selected Interviews and Other Writings, 1972–1977*, edited by Colin Gordon. Brighton, Sussex: Harvester Press.

Geddes, Andrew. 2015. "Temporary and Circular Migration in the Construction of European Migration Governance." *Cambridge Review of International Affairs* 28 (4): 571–588. https://doi.org/10.1080/09557571.2015.1018135.

German Federal Ministry of the Interior, Building and Community. 2020. *Virtual Meeting "Migration 4.0." Related to the German Presidency 2020.* https://migrationnetwork.un.org/migration-digitalization.

Hatton, Joshua. 2018. "MARS Attacks! A Cautionary Tale from the UK on the Relation between Migration and Refugee Studies (MARS) and Migration Control." *Journal for Critical Migration and Border Regime Studies* 4 (1): 103–132.

Ho, Justin Chun-Ting. 2020. "How Biased Is the Sample? Reverse Engineering the Ranking Algorithm of Facebook's Graph Application Programming Interface." *Big Data & Society* 7 (1): 205395172090587. https://doi.org/10.1177/2053951720905874.

Hughes, Christina, Emilio Zagheni, Guy J. Abel, Arkadiusz Wi'sniowski, Alessandro Sorichetta, Ingmar Weber, Andrew J. Tatem, et al. 2016. *Inferring Migrations, Traditional Methods and New Approaches Based on Mobile Phone, Social Media, and Other Big Data: Feasibility Study on Inferring (Labour) Mobility and Migration in the European Union from Big Data and Social Media Data.* Luxembourg: Publications Office.

International Organization for Migration. n.d. "Migrant." Key Migration Terms. Accessed March 15, 2021a. https://www.iom.int/key-migration-terms#Migrant.

International Organization for Migration. n.d. "Migration." Key Migration Terms. Accessed March 15, 2021b. https://www.iom.int/key-migration-terms#Migration.

Kikas, Riivo, Marlon Dumas, and Ando Saabas. 2015. "Explaining International Migration in the Skype Network: The Role of Social Network Features." In *Proceedings of the 1st ACM Workshop on Social Media World Sensors—SIdEWayS '15*, 17–22. Guzelyurt, Northern Cyprus: ACM Press.

Knowledge Centre on Migration and Demography, and Global Migration Data Analysis Centre. 2018. "Big Data for Migration Alliance Launch Event." June 25, 2018. https://knowledge4policy.ec.europa.eu/migration-demogr aphy/big-data-alternative-data-sources-migration_en#launch.

Law, John. 2004. *After Method: Mess in Social Science Research.* International Library of Sociology. London and New York: Routledge.

Law, John. 2008. "On Sociology and STS." *The Sociological Review* 56 (4): 623–649. https://doi.org/10.1111/j.1467-954X.2008.00808.x.

Law, John. 2012. "Collateral Realities." In *The Politics of Knowledge*, edited by Patrick Baert and Fernando Domínguez Rubio, 156–178. London and New York: Routledge.

Law, John, and John Urry. 2004. "Enacting the Social." *Economy and Society* 33 (3): 390–410. https://doi.org/10.1080/0308514042000225716.

Lu, X., L. Bengtsson, and P. Holme. 2012. "Predictability of Population Displacement after the 2010 Haiti Earthquake." *Proceedings of the National Academy of Sciences* 109 (29): 11576–11581. https://doi.org/10.1073/pnas.1203882109.

Marquez, Neal, Kiran Garimella, Ott Toomet, Ingmar G. Weber, and Emilio Zagheni. 2019. "Segregation and Sentiment: Estimating Refugee Segregation and Its Effects Using Digital Trace Data." In *Guide to Mobile Data Analytics in Refugee Scenarios*, edited by Albert Ali Salah, Alex Pentland, Bruno Lepri, and Emmanuel Letouzé, 265–282. Cham: Springer International Publishing. https://doi.org/10.1007/978-3-030-12554-7_14.

Messias, Johnnatan, Fabricio Benevenuto, Ingmar G. Weber, and Emilio Zagheni. 2016. "From Migration Corridors to Clusters: The Value of Google+ Data for Migration Studies." In *Proceedings of the 2016 IEEE/ACM International Conference on Advances in Social Networks Analysis and Mining: ASONAM 2016: San Francisco, CA, USA, August 18–21, 2016*, edited by Institute of Electrical and Electronics Engineers.

Mol, Annemarie. 1999. "Ontological Politics. A Word and Some Questions." Edited by John Law and John Hassard. *The Sociological Review. Sepcial Issue: Actor Network Theory and After* 47 (S1): 74–89.

Mol, Annemarie. 2002. *The Body Multiple: Ontology in Medical Practice.* Science and Cultural Theory. Durham: Duke University Press.

Nash, Sarah Louise. 2018. "Knowing Human Mobility in the Context of Climate Change. The Self-Perpetuating Circle of Research, Policy, and Knowledge Production." *Journal for Critical Migration and Border Regime Studies* 4 (1): 67–82.

Nieswand, Boris, and Heike Drotbohm. 2014. "Einleitung: Die Reflexive Wende in der Migrationsforschung." In *Kultur, Gesellschaft, Migration: Die Reflexive*

Wende in der Migrationsforschung, edited by Boris Nieswand and Heike Drotbohm, 1–37. Studien zur Migrations- und Integrationspolitik. Wiesbaden: Springer VS.

Palotti, Joao, Natalia Adler, Alfredo Morales-Guzman, Jeffrey Villaveces, Vedran Sekara, Manuel Garcia Herranz, Musa Al-Asad, and Ingmar Weber. 2020. "Monitoring of the Venezuelan Exodus through Facebook's Advertising Platform." Edited by Federico Botta. *PLOS ONE* 15 (2): e0229175. https:// doi.org/10.1371/journal.pone.0229175.

Plájás, Ildikó Z., Amade M'charek, and Huub van Baar. 2019. "Knowing 'the Roma': Visual Technologies of Sorting Populations and the Policing of Mobility in Europe." *Environment and Planning D: Society and Space* 37 (4): 589–605. https://doi.org/10.1177/0263775819837291.

Pollozek, Silvan, and Jan Hendrik Passoth. 2019. "Infrastructuring European Migration and Border Control: The Logistics of Registration and Identification at Moria Hotspot." *Environment and Planning D: Society and Space* 37 (4): 606–624. https://doi.org/10.1177/0263775819835819.

Raffnsøe, Sverre. 2008. "Qu'est-Ce Qu'un Dispositif? L'analytique Sociale de Michel Foucault." *Symposium. Canadian Journal of Continental Philosophy/Revue Canadienne de Philosophie Continentale* 12 (1): 44–66.

Rango, Marzia, and Michele Vespe. 2018. "Big Data and Alternative Data Sources on Migration: From Case-Studies to Policy Support. Summary Report. European Commission—Joint Research Centre (JRC), Ispra, Italy, 30 November 2017." https://bluehub.jrc.ec.europa.eu/bigdata4migration/ workshop-outcome.

van Reekum, Rogier. 2019. "Patrols, Records and Pictures: Demonstrations of Europe in the Midst of Migration's Crisis." *Environment and Planning D: Society and Space* 37 (4): 625–643. https://doi.org/10.1177/026377581 8792269.

Ruppert, Evelyn, and Stephan Scheel. 2019. "The Politics of Method: Taming the New, Making Data Official." *International Political Sociology* 13 (3): 233–252. https://doi.org/10.1093/ips/olz009.

Scheel, Stephan, Evelyn Ruppert, and Funda Ustek-Spilda. 2019. "Enacting Migration through Data Practices." *Environment and Planning D: Society and Space* 37 (4): 579–588. https://doi.org/10.1177/0263775819865791.

Scheel, Stephan, and Funda Ustek-Spilda. 2019. "The Politics of Expertise and Ignorance in the Field of Migration Management." *Environment and Planning D: Society and Space* 37 (4): 663–681. https://doi.org/10.1177/026 3775819843677.

Schultz, Susanne. 2019. "Demographic Futurity: How Statistical Assumption Politics Shape Immigration Policy Rationales in Germany." *Environment and Planning D: Society and Space* 37 (4): 644–662. https://doi.org/10.1177/ 0263775818772580.

Sîrbu, Alina, Gennady Andrienko, Natalia Andrienko, Chiara Boldrini, Marco Conti, Fosca Giannotti, Riccardo Guidotti, et al. 2020. "Human Migration: The Big Data Perspective." *International Journal of Data Science and Analytics*, March. https://doi.org/10.1007/s41060-020-00213-5.

Spyratos, Spyridon, Michele Vespe, Fabrizio Natale, Ingmar Weber, Emilio Zagheni, and Marzia Rango. 2018. *Migration Data Using Social Media: A European Perspective*. Luxembourg: Publications Office of the European Union.

Spyratos, Spyridon, Michele Vespe, Fabrizio Natale, Ingmar Weber, Emilio Zagheni, and Marzia Rango. 2019. "Quantifying International Human Mobility Patterns Using Facebook Network Data." Edited by Itzhak Benenson. *PLOS ONE* 14 (10): e0224134. https://doi.org/10.1371/journal.pone.0224134.

State, Bogdan, Mario Rodriguez, Dirk Helbing, and Emilio Zagheni. 2014. "Migration of Professionals to the U.S. Evidence from LinkedIn Data." In *SocInfo 2014*, edited by Luca Maria Aiello and Daniel McFarland, 531–543. Lecture Notes in Computer Science 8851. Springer International Publishing.

State, Bogdan, Ingmar Weber, and Emilio Zagheni. 2013. "Studying International Mobility through IP Geolocation." In *Proceedings of the Sixth ACM International Conference on Web Search and Data Mining, WSDM 2013, February 04–08, 2013, Rome, Italy*, edited by ACM International Conference on Web Search and Data Mining. New York: ACM.

Stewart, Ian, René D. Flores, Timothy Riffe, Ingmar Weber, and Emilio Zagheni. 2019. "Rock, Rap, or Reggaeton? Assessing Mexican Immigrants' Cultural Assimilation Using Facebook Data." In *The World Wide Web Conference—WWW '19*, 3258–3264. San Francisco, CA, USA: ACM Press.

Stielike, Laura. 2017. *Entwicklung Durch Migration? Eine Postkoloniale Dispositivanalyse am Beispiel Kamerun-Deutschland*. Kultur und soziale Praxis. Bielefeld: Transcript.

Taylor, Linnet. 2016. "No Place to Hide? The Ethics and Analytics of Tracking Mobility Using Mobile Phone Data." *Environment and Planning D: Society and Space* 34 (2): 319–336. https://doi.org/10.1177/0263775815608851.

Taylor, Linnet, and Lina Dencik. 2020. "Constructing Commercial Data Ethics." *Technology and Regulation*, 1–10. https://doi.org/10.26116/TECHREG.2020.001.

Taylor, Linnet, and Fran Meissner. 2020. "A Crisis of Opportunity: Market-Making, Big Data, and the Consolidation of Migration as Risk." *Antipode* 52 (1): 270–90. https://doi.org/10.1111/anti.12583.

UN DESA. 1998. "Recommendations on Statistics of International Migration, Revision 1." https://unstats.un.org/unsd/publication/SeriesM/SeriesM_58rev1e.pdf.

UN Global Pulse, and UNHCR Innovation Service. 2017. "Social Media and Forced Displacement: Big Data Analytics & Machine-Learning." https://www.unhcr.org/innovation/wp-content/uploads/2017/09/FINAL-White-Paper.pdf.

UN Statistics Division. 2017. "Improving Migration Data in the Context of the 2030 Agenda." https://unstats.un.org/unsd/demographic-social/meetings/2017/new-york--egm-migration-data/Background%20paper.pdf.

United Nations. 2018. "Global Compact for Safe, Orderly and Regular Migration. Resolution Adopted by the General Assembly on 19 December 2018."

United Nations Expert Group Meeting. 2017. "Improving Migration Data in the Context of the 2030 Agenda. Recommendations, New York Headquarters, 20–22 June 2017." https://migrationdataportal.org/sdgs#0.

Will, Anne-Kathrin. 2019. "The German Statistical Category 'Migration Background': Historical Roots, Revisions and Shortcomings." *Ethnicities* 19 (3): 535–557. https://doi.org/10.1177/1468796819833437.

Yeung, Karen, and Martin Lodge, eds. 2019. *Algorithmic Regulation*. 1st ed. Oxford University Press. https://doi.org/10.1093/oso/9780198838494.001.0001.

Zagheni, Emilio, Venkata Rama Kiran Garimella, Ingmar Weber, and Bogdan State. 2014. "Inferring International and Internal Migration Patterns from Twitter Data." In *WWW '14 Companion: Proceedings of the 23rd International Conference on World Wide Web: April 7–11, 2014, Seoul, Korea*, edited by Chin-Wan Chung, International WWW Conference, Association for Computing Machinery, Hypermedia and Web Special Interest Group on Hypertext, and International World Wide Web Conference Committee.

Zagheni, Emilio, and Ingmar Weber. 2012. "You Are Where You E-Mail: Using E-Mail Data to Estimate International Migration Rates." In *WebSci'12: Proceedings of the 3rd Annual ACM Web Science Conference, 2012 : Evanston, IL, USA*, edited by Web Science 2012, Brian Uzzi, Michael Macy, Noshir Contractor, and ACM Digital Library.

Zagheni, Emilio, Ingmar Weber, and Krishna Gummadi. 2017. "Leveraging Facebook's Advertising Platform to Monitor Stocks of Migrants." *Population and Development Review* 43 (4): 721–34. https://doi.org/10.1111/padr.12102.

PART II

Ethical Challenges in Digital Migration Research and Beyond

Impossible Research? Ethical Challenges in the (Digital) Study of Deportable Populations Within the European Border Regime

Leandros Fischer and Martin Bak Jørgensen

INTRODUCTION

Some time ago, during our DIGINAUTS project, one of the authors of this contribution attended an important migration conference on the digitalisation of bordering practices. His presentation dealt with the digital strategies of those potentially facing deportation. After elaborating on some basic findings of our research among precarious migrants in Hamburg and northern Denmark, he proceeded to show some translated comments and postings from a Facebook group, here called platform.

L. Fischer
Hamburg, Germany
e-mail: fischer@hum.aau.dk

M. B. Jørgensen (✉)
Department for Culture and Learning, Aalborg University, Aalborg, Denmark
e-mail: martinjo@hum.aau.dk

© The Author(s) 2022   141
M. Sandberg et al. (eds.), *Research Methodologies and Ethical Challenges in Digital Migration Studies*, Approaches to Social Inequality and Difference, https://doi.org/10.1007/978-3-030-81226-3_6

This was a closed, Arabic-language group of about 330,000 members, which was mentioned during one of the qualitative interviews for our research. Our project assistant, an Arabic native speaker, requested access to the group, which was immediately granted. She then selected some posts and comments, anonymised them, and created a folder for the project. The co-author presented these comments as an example of various migrant strategies that not only pointed to linear journeys towards preferred destinations but were indicative of multiple sojourns in different European countries due to long waiting periods of application processing or rejection, as well as return journeys. In line with our approach—grounded largely on the Autonomy of Migration (AoM) concept—the co-author intended to highlight the creativity and ingenuity of migrants who—despite all odds—managed to find ways to circumvent restrictive bordering practices. It was an effort to emphasise agency, rather than victimhood.

Nevertheless, the audience was not overly receptive to this point of view. During the discussion, methodological questions arose. One participant in particular asked whether our student assistant had notified the group members that she was conducting research within the closed group. The co-author replied that such a thing would be of no use in a group of 330,000, where queries are deleted as soon as they are adequately answered. Instead of satisfying the discussion participant, this answer only seemed to agitate him, as well as other people in the room. "How do you know that your data are not being collected by FRONTEX [the European agency tasked with guarding the continent's militarized borders]?" The co-author replied that, given the ease with which access to the group could be gained, FRONTEX would already have placed informants within the platform. The co-author sensed that *he* was not personally satisfied with his answer. Indeed, how could we be sure that our data—indicating ways of circumventing border restrictions—would not ultimately be used against migrants? The discussion ended with no conclusion.

This contribution is one attempt to answer some pressing methodological as well as ethical dilemmas of digital research with vulnerable groups. How does one conduct research that aims to highlight the agency of migrants without inadvertently placing them in danger? How is the question of inherently uneven power differentials played out in this case? And doesn't the overwhelming majority of social science fieldwork and qualitative research involve unequal power relations almost *by definition*? There are certainly more papers published on refugees, urban slum dwellers, or oppressed minorities than, let's say, on politicians, representatives of the finance industry, or top-brass military officers. This is partly because the former group is infinitely more approachable than the latter, who is concealed in gated communities, high-rise luxury apartments, or military bases.

This chapter departs from our work within the DIGINAUTS project (see Sandberg and Rossi, Introductory chapter to this volume). The purpose of this project was to investigate how information and communications technology (ICT), the digital practices of migrants as well as of aiding organisations and initiatives of the receiving countries intermediate and constitute new sociotechnical networks of community and solidarity, in turn re-enacting migrants as political subjects in/of the European border regime. As a subproject, the two authors of this chapter were involved in and focused on the Danish–German borderland. In 2015, more than 21,000 people applied for asylum in Denmark (Udlændinge-, Integration- og Boligministeriet 2016). Many used Denmark as a pathway to reach Sweden before border controls were enforced. In 2015, Germany received over one million claims for asylum and large numbers passed through the country to get further north. Hamburg has long been a hub for especially Sub-Saharan migrants coming to Hamburg via Libya and Lampedusa (Jørgensen 2019).

Digital connectivity has been important for both the migrants staying in Germany trying to find information on rights, housing, work, legal help, and civil society groups and aid workers, and for the migrants seeking to enter Scandinavia. During the subproject, we conducted fieldwork in Hamburg and various places in Denmark to compare patterns in ICT use and survival strategies between the latter group and the recent incoming groups—especially from Syria.

Very soon after initiating the project and beginning work on the subproject, we faced a new situation. In several European countries today, we can identify a paradigm shift from migrant integration to the securitisation of migration and deportation. In Denmark, the government has generally stepped-up deportations and we are witnessing an expansion of the category of deportable populations. This politicisation of immigration in Denmark has caused enormous insecurity among migrants. Similar tendencies can be observed in Germany, where the end of political *Willkommenskultur* by the state has been met with an increase in deportations to Afghanistan and the Balkans. This means that potential deportees and undocumented migrants must increasingly develop survival strategies for if and when they are deported. Furthermore, migrants are starting to share information and strategies online on how to return to Turkey and the Middle East. In the case of Denmark, we are seeing a still increasing number of people disappearing from the authorities' radar, going to other European countries and living as irregular migrants or attempting to apply for asylum through loopholes in the Dublin agreement. This development led us analytically towards a focus on anti-deportation. The change of focus also developed into methodological and ethical issues and a self-investigation on how to work on these issues.

Departing from the issues outlined so far, this chapter asks how we (as researchers) collect data, how we engage with our informants, how we disseminate our results, and what we seek to achieve by and through our research. The chapter is structured as follows. Firstly, we discuss migration research ethics and the various challenges we identify here. We use this discussion to reflect on the polarising effects of politicised research and how to take a stance. Secondly, we engage in a critical debate with the principle of "doing no harm" and ask if it is possible to outline a research position with the opposite goal, i.e. making the repressive and destructive features of the border regime visible. Here, we highlight two different approaches: the militant research approach and the AoM approach. In short, militant research is a politically engaged research practice that seeks to be capable of articulating involvement and thought. AoM can be best described as an attempt to theorise the role of migrant agency in the constitution of contemporary border regimes (Fischer and Jørgensen, forthcoming a). Its emphasis is placed on the primacy of movement over control (Bojadžijev and Karakayali 2010; Karakayali and Tsianos 2010; Mezzadra 2011), as well as the development of socialities and mundane practices independent of sovereign control among people on the move, dubbed the "mobile commons" (Fischer and Jørgensen, forthcoming a; Papadopoulos and Tsianos 2013). The AoM the approach "seeks to reinterpret the effects of seeing regular, irregular, transit and other forms of migration as constitutive factors of border policies, architectures, and practices" (Casas-Cortes et al. 2015, 897). Or in simpler terms and conclusively, the approach makes mobility and migration the starting point of analyses and conceptualises migrants as having agency (Agustín and Jørgensen 2019). In our work on DIGINAUTS, the militant research perspective has influenced our methodological grounding and the AoM has guided our conceptual and theoretical perspectives (see also Galis in Chapter 7 of this volume). Thus, the third section of this chapter engages with the militant research approach and discusses how and if the departure from a politically committed approach also offers an ethical research strategy, i.e. politically engaged *as* ethics. Fourthly, we bring in the AoM approach and, along similar lines, discuss how the principles of AoM influence our methodology and intentions with our research. In the final part of the chapter, we combine these perspectives and situate our own research studies in a final discussion.

Migration Research Ethics

In the migration research literature, ethical issues are very often brought to the forefront. As stated by other scholars (e.g. Yalaz and Zapata-Barrero 2018; Zapata-Barrero and Yalaz 2020) the research context is key here, as qualitative migration studies very often involve being in contact with people whose migratory experiences can have very different characteristics. There is a world of difference between studies of Western retirees in Indonesia (Bell 2017), Swedish retirement migrants living in precarious conditions in Spain (Gavanas and Calzada 2016), the privileged mobilities of expats residing in Paris, Singapore, and Bangalore (Polson 2016) and studies on migrants with trajectories characterised by traumas, abuse, and even torture (Nimführ and Sesay 2019; Van Liempt and Bilger 2018). All such studies require ethical research virtues. However, the differences and the stakes that are embedded in these studies force us to ask where to draw the ethical lines between risking interlocutors' safety and pursuing our own research ends, in direct contrast, for instance, to medical research, where such issues are keenly debated (Düvell et al. 2010). Despite the necessity to explicitly draw this line—in migration studies—Düvell et al. have argued that this rarely happens (2010). Picking up on the claim made by Düvell et al., Zapata-Barrero and Yalaz (2020, 2) formulate the basic questions we need to ask:

> These questions now need to be considered as key-issues belonging to the same research design process: How do we ethically carry QR [qualitative research] with migrants? How do we solve particular ethical situations and dilemmas? How do we identify and manage ethical risks in conducting QMR [qualitative migration research]? What has to be the reference framework for assessing ethical risks? Do these ethical considerations affect the quality and objectivity of the research? Are universal ethical codes of conduct applied to QMR enough for dealing with particular situations?

These questions are also addressed in the literature (e.g. Birnie 2019; Rodgers 2004; Siapera and Creta 2020; Van Liempt and Bilger 2009, 2018). The questions call for both an ethical awareness and an ethical reflexivity in regards to how we (as researchers) collect data, how we engage with our informants, how we disseminate our results, and what we seek to achieve by and through our research. Again, none of the reflections that we raise here are novel when considered in isolation. These are issues with a long academic trajectory and debate within ethnology, ethnography, and social anthropology. Within anthropology and ethnography, these issues are part of a reflexive turn (e.g. Clifford and Marcus 1986; Foley 2002). This turn has emphasised how the person of the anthropologist can affect the ethnographies they write and has forced anthropologists to tell the story of their integration and interactions within the community they were studying (e.g. Venkatesh 2013). This kind of awareness and reflexivity can be broken down into various issues.

Firstly, as already mentioned, there is a categorical difference in the kind of ethnical reflexivity needed when studying migrants who have voluntarily entered migration (can be both privileged and non-privileged migratory processes and statuses) and those who have entered the migratory process in vulnerable and precarious circumstances. The latter is especially true when we contemplate the field of migration and include illegalised migrants without recognised papers, people living in camps, racialised minorities, rejected asylum seekers, people facing deportation, having been trafficked, and similar vulnerable and precarious situations. Here, we need to acknowledge and account for inequalities and power dynamics. The problem is, however, that these categories of migrant statuses are not self-evident and do not give us a fixed answer to the question of which migrant groups to work with and how. Categories of difference have a crucial position in academic research as well as policy-making. They serve to distinguish and differentiate between groups in society. They can appear in the form of crude dichotomies or in complex and sophisticated forms resting on constructivist and intersectionalist perspectives. Nevertheless, using categories of difference also causes something to exist and there may be implications through the particular application of specific categories (Jørgensen 2012). Put more simply, categories and their definitions matter. Categories of race, ethnicity, gender, or social divisions are all categories of difference, which serve to position the particular individual or group in a social and economic location. Similarly, what is common for most of the different academic approaches—despite disciplinary demarcations and intentions of wanting to analyse discrimination, marginalisation, or inequalities, for instance—is that in order to study immigrants and ethnic minority communities, the focus necessarily ends up highlighting differences. In order to enter this research field, studies inevitably reproduce various distinctions between "us" and "them". Academic approaches "write differences", so to speak, into the texts they produce (Nayak 2006; Rosenblum and Travis 2008). Research may itself construct categories of difference, for instance, by racialising research (Ali 2006).

As also addressed by Zapata-Barrero and Yalaz, this difference in what and whom we study relates to a second debate between the quality and objectivity of the research and thinking ethically (2020, 2). In the literature, thinking ethically has been discussed through three ethical universal provisos: do no harm; respect autonomy; and ensure equitable sharing of benefits (ibid.; Flick 2018; Krause 2017). Likewise, these ethical provisos are not adequate for describing the participatory action-research approaches that aim at intervening and spurring social change. We will return to this part of the discussion below, however, an initial argument here is that migration research is a normative activity that refers to conscious social and political engagement. This engagement must be able to deal with both potential harms and benefits (Düvell et al. 2010).

Thirdly, there is the issue of how we disseminate research. The ambition of producing knowledge that can be used by social actors, including migrants themselves, to reveal exploitation and repression and improve their conditions is also put at risk of being abused by anti-immigrant forces, states, and security agents. Our initial vignette is a good example of this dilemma. We can find attempts to develop best practices in the literature. In their work on irregular migration, the aim of Düvell et al. has: "Not been just to produce a 'body of knowledge' but to address the misperceptions and misconceptions surrounding irregular migration, and to minimise the risks and maximise the benefits, firstly for the researched group and, secondly, for other stakeholders" (2010, 228–229). They continue later in the article by raising a number of both hypothetical and practical questions:

> Should all or only some results be published? Who is the audience? How will our results be received and discussed at a given time period (in the light of related political and public debates) and how may they be (ab)used? The question also arose as to what extent we can control and influence the (ab)use of our findings. (ibid., 235)

Having worked on irregular migration ourselves, we recognise these challenges and dilemmas. One of the co-authors of this chapter ended up experiencing the political implications of producing and disseminating research knowledge. Following the publication of an edited volume on irregular migration in Scandinavia (Thomsen et al. 2010), the immigrant-critical Danish People's Party called for increased control as a consequence of the research results. Emphasising that there is no exact information about the number of irregular migrants does not help, as numbers rapidly gain an existence of their own. The development in Denmark at the time provided an interesting example.

Before 2012, the Danish police estimated that there were around 5000 irregular migrants in Denmark, but in 2012 the estimate soared to between 20,000 and 50,000 irregular migrants in the country (Glerup 2012). The estimate was not based on any research project or new methodology but on three interviews with anonymous sources within the police (Glerup 2012). Nevertheless, this number was reproduced by the national public and private media (e.g. Fahrendorff 2012) and ultimately became part of the Danish People's Party's 1st of May Campaign (in 2012)—"Do something about the large number" (40,000) of irregular migrants they advocated for (see Thomsen and Jørgensen 2012). An unintended consequence was that our research was used to legitimise a call for restrictions and penalise practices. Disseminating research results not only includes the dilemma of making sensitive knowledge available for security agents and the police and penal systems but also the risk of being politicised towards aims over which the researcher has no control.

Fourthly, we must consider the aims and purposes of our research—and if these contribute to social and political change towards equality and social justice. When it comes to qualitative ethnographic work, there has been a long debate regarding the ethics of research with vulnerable groups. Migratory experiences characterised by vulnerability and precarity fall under this discussion. Ultimately, these are not only questions of ethics but an epistemological question about the choice and standpoint of the researcher in regards to the purpose and value of knowledge. In the article "'Stop Stealing Our Stories': The Ethics of Research with Vulnerable Groups", Pittaway et al. (2010) discuss some challenges and opportunities facing those working to integrate participatory methods into human rights-based research. Their article draws on refugees' experience to analyse how interviewees perceive the interaction with and (lack of) trust in researchers, emphasising three interdependent issues: risks, distrust, and hierarchical distances. They discuss an approach they have designed for working with communities and individuals experiencing multiple oppressions and persecution. This approach reflects the principles of anti-oppressive social work and the ethics of undertaking research with vulnerable populations (ibid.). The relationship between the researcher and people in vulnerable positions, in our case migrants at risk of deportation, is definitely unequal. As Düvell et al. write about their work with irregular migrants, such persons, despite the asymmetrical relationship, nevertheless hold a position of key influence in the research context, namely the ability to decide whether or not to tell their stories and participate in the research (2010, 233). This holds some truth but as Düvell et al. also explain, informants can easily misinterpret the role and potential influence of the researcher and choose to talk with researchers, believing those researchers have the power to improve their position. We recognise such relationships and encounters in our own work. One of the co-authors visited an informant at a deportation centre, not to conduct interviews but to follow up on the well-being and situation of that person. However, rumours immediately spread that he was there to help people with their asylum cases and a number of other women at the centre lined up to tell their stories in the hope of receiving help. Although, in an unpredicted way, such a situation can end up generating important data, none of these conversations were recorded or used for research. Here, the situation was not a matter of stealing people's stories but a matter of people wanting their stories to be actually heard, as the women felt that none of the staff or migration authorities had listened to them without misinterpreting what they were saying. They initially wanted to share their stories with a clear aim though—that talking to the co-author could change their asylum rejection. This is just one example of many comparable situations we have encountered, both within the DIGINAUTS project but also in previous work. It shows how, during practical work, the relatively abstract academic aims of a research project can be confronted with the lives being lived by the people we engage with. In practice, we are met with real people struggling individually and collectively who—despite the asymmetrical relationship—meet us, the researchers, with a claim that their struggles should not only be understood but that we should stand beside them and offer the help they expect us to provide to improve their situation. This represents another ethical challenge. Wissink describes this dilemma well as a gap between ethics in the phase of research design, which may be mainly inspired by "text-book ethics" and the ethical dilemmas a researcher may experience in the field (2019, 2) (for a discussion on the emotional implications of conducting research in insecure places among migrants in extremely difficult situations, see Nyberg Sørensen in Chapter 8 of this volume).

The four issues we have so far addressed in the discussion: how we (as researchers) collect data, how we engage with our informants, how we disseminate our results, and what we seek to achieve by and through our research, are all relevant, and we can evidently refer to a theme that traces back far into the literature. However, what we faced in our work were not only concerns about how to work with deportable populations but also why we were doing so. As researchers engaged in studies on sensitive and politicised issues, we often cannot avoid prompting a polarising effect. Düvell et al. point this out and continue by stating that "sometimes, our findings have been of such an explicit nature that it has proved difficult not to take sides" (2010, 236). Is taking sides or taking a stand a problem though? And for whom? Situating our research by drawing on militant research approaches and AoM, is obviously not a problem for the authors of this chapter, but it is nevertheless a question flagged in literature and one that deserves an answer. Düvell et al. themselves answer this rhetorical question by formulating a position:

> Researchers, however, are not primarily advocates or social workers but academics, and they are subject to a set of complex responsibilities for high quality and ethical research. They have responsibilities towards their subjects, their profession, their funding bodies and society at large. In our case this meant we had to negotiate a balanced attitude between contrasting perspectives and opposing aims and interests. (ibid.)

For Düvell and Triandafyllidou, the path has been to pursue advocacy from an NGO platform rather than from academia and engage in a kind of public sociology (cf. Burawoy 2005). We can follow the position outlined here and likewise the personal stances taken by those authors. At the same time, this discussion opens up a range of new questions. Is there the scope for a critical or activist engagement within academic knowledge production? How can the effects of crushing machinery such as the FRONTEX and EU border regimes with their increased militarisation be described and analysed while at the same time avoiding revealing tactics and strategies used to circumvent and transgress borders? How can we respect and acknowledge vulnerability while at the same time wanting to avoid victim-isation and take the agency of migrants seriously? Is it possible to adhere to some kind of militant research ethics and what are the costs for the people we engage with who are in a different position to ourselves? Is it possible to write about such issues in a way that does not harm our infor-mants but similarly does not harm an abusive, repressive, exclusionary, and dehumanising migration system that is constructing and controlling deportable populations? In the next part of the chapter, we discuss the possibility of such an approach that situates research ethics as part of research politics and we identify our own position as researchers.

FROM "DO NO HARM" TO "DO A LOT OF HARM"—TOWARDS POLITICS OF CRITICAL ETHICS

In their protest anthem "Killing in the Name" from 1992, the US alter-native rock band Rage Against the Machine takes a stance against racism and abuse of power and calls for a revolution by repeating the line "Fuck you, I won't do what you tell me" over and over. Anyone old enough to have been to one of their shows or been on the dance floor when the song was played will recognise the urge to destroy the world around you and create an alternative order. Having worked with people who have experi-enced the effects of the European border regimes (cf. Stierl 2018) almost inevitably makes you want to take sides and engage in research practices that could help confront and challenge this system. However, destroying the system is not a real option and not a purpose that will bring home much funding from universities or external funders. Moving on from this playfully deliberate provocative stance, we can discuss how it is possible to move towards an engaged and critical research position.

Here, we return first to the ethical dictum of "do no harm". Often considered the golden rule of research ethics, it is also considered insuf-ficient by migration researchers (e.g. Block et al. 2013; Pittaway et al. 2010; Zapata-Barrero and Yalaz 2020). Writing about the ethics of media research with refugees, Siapera and Creta unfold this argument:

> Ethical positions alone are not adequately equipped to address the prob-lems that refugees in detention are facing. For us, the authors of this chapter, occupying the positions of media workers, activists and researchers at the same time involves contradictions that we are unable to address in purely ethical and moral terms, by invoking vague ethical principles such as 'do no harm' or 'protect vulnerable people'. (2020, 236)

The problem is that merely following "procedural no-harm procedures" (e.g. blindly following formal procedures of confidentiality and privacy)—as Zapata-Barrero and Yalaz (2020, 6) formulate it—can still cause harm to sensitive populations if researchers do not critically evaluate the rising ethical issues. Secondly, this approach does not guarantee or contribute to enabling participation of vulnerable and hard-to-reach populations. They rightly—in our opinion—claim that migration researchers not only have a duty to protect migrants but also to empower them (ibid.). A viable strategy is perhaps not to listen to "Killing in the Name" but to revisit the 1976 Fleetwood Mac classic "Go Your Own Way". Is it possible to develop an ethical research position that both aims at doing no harm to the people engaged in the study and, at the same time, working for social and political change and ultimately standing in solidarity with the people at risk of deportation? Jacobsen and Landau (2003) quote Turton for contending that, "researching other people's suffering can only be justified if the research explicitly aims at alleviating that suffering" (Turton 1996, 96, in Jacobsen and Landau 2003).

In a recently published article, Stierl makes the case for an engaged scholarship that does not shy away from intervening in the contested field of migration with the intention not to fix but to amplify the epistemic and other crises of the European border regime (2020). In the article, Stierl argues that Migration Studies tends to reify and fetishise epistemic objects such as "migration" and "migrants" (here quoting De Genova et al. 2018, 257) (ibid., 8). Stierl also contends that, for instance, cartographic representations such as those constructed by FRONTEX, are not neutral but have concrete human consequences (see also the work of Van Houtum and Lacy 2020) and likewise, that migration statistics (like maps) account for the subjective experiences of borders (Stierl 2020, 10).

Stierl criticises the "do no harm" principle as being inadequate and asks if it should not also be expanded to engagements with policymakers. Stierl ends by discussing epistemic interventions that can produce counter-empirics in order to expose the violence of the EU border regime. In his concluding reflections, he discusses the impact of an activist engagement as a way to produce critical knowledge on migration and concludes that:

Maybe the 'do no harm' principle needs to not merely be expanded to include engagements with the makers of migration policies, it may need to be reversed. *Do harm* could be the motto for a critical and impactful scholarship of migration that locates, and expands, ruptures in the EUropean border regime. (ibid., 16; italics in original)

Is it possible to take a research position here that actually seeks to *do harm* (moving from *Fleetwood Mac* back to *Rage Against the Machine*), while, at the same time, respecting the credos of (still) doing no harm to the people we work with, respecting autonomy, and ensuring equitable sharing of benefits? In the next sections, we describe and discuss two different research approaches, namely a militant research/militant anthropology approach and the AoM approach and discuss the kind of ethical positioning these two approaches offer.

MILITANT RESEARCH AS AN ETHICAL RESEARCH STRATEGY

Previously, we discussed if taking sides in a field of politicised research constitutes a methodological and/or ethical problem. From a militant research perspective, this is a futile discussion. The starting point for militant research is not an academic researcher seeking to further a particular strand of knowledge, but the context of political struggle itself (Halvorsen 2015). Militant research "is an intensification and deepening of the political", claim Shukaitis and Graeber (2007, 9). As an approach, militant research sees research and activism as co-constituted and is oriented solely "by invested militant activists for the purpose of clarifying and amplifying struggle" (Team Colors Collective 2010, 3). Militant research connects to other engaged and militant approaches within anthropology, ethnography, and sociology (see Jørgensen 2019). Scheper-Hughes' seminal work in 1995 called for a "militant anthropology" and the "primacy of the ethical", and for anthropologists to become morally and politically engaged. Juris coined the notion of "militant ethnography" to describe this approach. He depicts this as "developing a model of politically committed ethnographic research that uses engaged ethnography as a way to contribute to movement goals while using my embedded ethnographic position to generate knowledge of movement practices and dynamics" (Juris n.d.; see also 2007, 2014). Mathers and Novelli call for an engaged ethnography (2007). These positions originate in different disciplines but pursue the same goal: solidarity with the research subjects and a research praxis that produces knowledge on how micro-processes of resistance are linked to macro-processes of repression (be it against neoliberal globalisation or the border regimes) (Mathers and Novelli 2007). This is both a political and ethical stance. A militant research position in this way highlights engagement, the priority of the ethical (as in committed research), possible interventions and disruptions in the field we study, and solidarity between citizens.

We have continued this kind of political engagement in our current research project. Understanding migrants' practices against deportation (Fischer and Jørgensen, forthcoming b) also necessitates an understanding of the EU border regimes and their effects. We have combined our ethnographic approach with a methodology from political ethnography. The latter makes it possible to interpret the legal and policy framework and to capture dynamics and relations beyond and outside the policy frameworks themselves and is becoming an increasingly popular approach for understanding politics and government (Boswell et al. 2019). From a policy-analysis perspective, it opens a window into the micro aspects of politics (Kumar 2014). The methodological aim here is not to identify causal inference (a common aim in mainstream political science) but rather interpretation and edification. Pursuing the militant research approach is not equal to conducting some kind of vigilante crusade against the oppressors or sailing under a black flag in all our academic and personal doings—we are not claiming to save the world but *insist* on engaging in critical research that is not objective but politically committed. Mathers and Novelli pick up on Scheper-Hughes and argue: "[T]he ethnographer [here broadly researcher] may find many paths to ethical and political commitment, but each of them involves him/her in undertaking a variety of acts of solidarity" (2007, 245).

Solidarity is also an ethical commitment that entails concrete (research) practices. In our work on the DIGINAUTS project, we followed the standard ethical guidelines and have done what we could to ensure informed consent and avoid any risk of harm to our informants. This became especially important when interlocutors were telling us about strategies on how to avoid deportation or tactics used to re-apply for asylum elsewhere. We have also engaged with our informants and changed our research focus towards what was important to them (as we started to understand the hardship faced by people at risk of deportation) rather than maintaining the focus on the trajectory from outside Europe to the German–Danish borderland and tried to produce knowledge that emphasises their struggles. In practice, we have used very open questions, allowing the interlocutors to be in control of what was important, which stories to tell and in defining what it means to be at risk of deportation. They have told us what to look at, which directions we should pursue, what was not relevant, and what they wanted to happen. We have not tried to (and would not have been able to) change the outcomes of their claims for asylum. However, we have connected people with other people in similar conditions and people in solidarity. We have told them of physical and digital resources (such as legal aid and networks helping people facing deportation) and passed on other people's experiences. Like the people we have met, we have engaged in political struggles, in our own way—through writing and dissemination. We do not claim to be in a comparable position, as none of the authors were or are in danger of deportation, but we took sides and stand alongside people in struggle. For us, this signals a move from ethics as a specific research practice based on ethical provisos to a political position where research is situated in politics and based on an imperative to act—one way or another.

Research has the purpose of making repressive and unjust structures visible. Instead of focusing attention solely on the individual stories and micro-forms of resistance, one aim can be to reveal the structures that create repression and destitution (also following Stierl). When rejected asylum seekers at the deportation centre Kærshovedgaard in Denmark began a hunger strike to draw attention to their suffering, we supported their struggles through our writing and the privileged platform to which academics have access. The movement, initiated by refugees, aimed to publicise the consequences of the politics of dehumanisation; politics that "kill slowly" and that are structurally produced and legitimised by law. Our, self-proclaimed, role here was to conduct a political analysis of the system producing these effects and communicate the experiences of the people trapped in this system (e.g. as in Jørgensen et al. 2018). Two years later, the situation had not been improved and people residing at Kærshovedgaard asked us again to help share their stories with the public.

Autonomy of Migration and Research Ethics

In our work on migrants' physical and digital strategies against deportation, we draw on an approach stressing autonomy, namely the AoM approach. In the book *Border as Method*, Mezzadra and Neilson (2013) go on to demonstrate how the proliferation, mobility, and deep meta-morphosis of borders are key features of "actually existing" processes of globalisation. Their book builds on the AoM approach (e.g. Bojadžijev and Karakayali 2010; De Genova 2017; Papadopoulos and Tsianos 2013) and they draw on these ideas "to frame the border epistemologically and methodologically in order to develop a conjunctural analysis of current capitalist configurations" (Casas-Cortes et al. 2015, 897). This links the reading of borders to multi-scalar processes of political geography (ibid.). The AoM approach makes mobility and migration the starting point of analyses and conceptualises migrants as having agency. In this way, borders follow migration and not the other way around by constituting collective action that challenges institutional power to reshape the border regime (Mezzadra 2011). Migration here is akin to a social movement.

Our investigation of anti-deportation strategies is based on elements from AoM, and specifically the concept of the "mobile commons" (Papadopoulos and Tsianos 2013; Trimikliniotis et al. 2016). For our analyses, the important aspect of these approaches is that they allow us to understand how people on the move act without ending in victim-isation. Reading anti-deportation strategies with an AoM perspective allows us to identify and understand social and political spaces created by refugees facing deportation. Instead, migration is theorised as linked to the agency of migrants themselves, specifically the desire for the freedom of movement. In addition, the desire for movement should not be simply conceptualised as merely the need for social mobility, but as one also moti-vated by a condition of "stuckedness" and a lack of "existential mobility", the "sense that someone is going somewhere in life" (Hage 2009). The claims made by the people we met during our project denotes both a right to receive protection as well as the right to decide where to receive this protection; migrants do not simply desire the right to stay but—like those enjoying full civil rightswant to decide themselves where their place of settlement should be. As we write in a forthcoming chapter for another volume, this means that often migrants risk a secure existence in one state to live elsewhere, motivated by a sense of justice (Fischer and Jørgensen, forthcoming b).

The concept of mobile commons describes an infrastructure always in the making, and that encapsulates the innumerable uncoordinated but cooperative actions of mobile people who contribute to its making. People on the move "create a world of knowledge, of information, of tricks for survival, of mutual care, of social relations, of services exchange, of solidarity and sociability that can be shared and used and where people contribute to sustain and expand it" (Papadopoulos and Tsianos 2013, 190). Digital "infrastructures of connectivity" (ibid.) constitute a basic component of the mobile commons, providing those on the move with useful information, whether in the form of navigation apps, social media, online fora, or simply as "hardware", such as battery-loading docks for smartphones present along the Balkan route prior to its closure.

So far, we have underlined the normative and epistemological under-pinning of the AoM approach.

For us, it has also carved out research ethics on how and why we conduct research. Here, the term ethics implies doing research that stands on the side of migrants. Papadopoulos and Tsianos also underline the importance of this axiom:

> In fact, the autonomy of migration approach is only possible if it contributes to creating conditions of thick everyday performative and practical justice so that everyday mobility, clandestine or open, becomes possible. This is a form of thick justice which creates new forms of life that sustain migrants' ordinary movements. (Papadopoulos and Tsianos 2013, 192)

Analysing and understanding the mobile commons of deportable populations from one perspective can seem to clash with the (alleged) vulnerability of irregular migrants as mobile commons point to agency and spaces of emancipation (cf. Düvell et al. 2010). However, this assumption is incorrect. The aim of our approach has not been to map an existing infrastructure but to show how people in a precarious position are also not without a voice (as the hunger-strike example illustrates). Understanding physical and digital strategies against deportation helps produce counter-knowledge that becomes part of the mobile commons. Here, we are not assuming that people on the move or people facing deportation will go to academic journals, but knowledge on cracks, fric-tions, resistance, and solidarity also travels along other trajectories. Other people within academia engaged in migrant struggles will have access to knowledge. In its own way, research knowledge can also become part of the commons. Research knowledge is here produced with the aim of interfering with and spurring social and political change. It is engaged scholarship produced as a type of politics articulated by both deportees and researchers by different means.

How to Study Deportable Populations—Impossible Research or Political Research

Our work on studying deportable populations within the EU border regime has led us to work with different types of data and methodologies. We have both pursued ethnographic on-site fieldwork as well as looking into digital strategies. The ethical challenges of digital research revolve around the disconnectedness between identity and bodies and are fraught with both danger (surveillance) and opportunity/agency (mobile commons, AoM). This challenges researchers: not just to protect and do no harm but also to realise that the internet has different rules.

One of the insights we gained from the project was that the better and more protected the social position of migrants (by civil society) (e.g. Germany), the more complicated it was to gain access to them (arrangements with churches, building trust). Researchers should embrace or at least accept potential hurdles that can make it more difficult/slower to find the necessary interlocutors and encourage them to talk, due to the fact that the research subjects in question are better protected and more aware of their rights. This contrasts, for example, with refugee camps or makeshift transit locations in peripheral countries (e.g. Greece), where researchers may have easier access to vulnerable populations as their presence is very visible—although it does not follow from this that anyone residing there would want to talk to researchers doing fieldwork nor that the government would allow researchers access (on access see Rozakou 2019). Obviously, trying to access either group demands ethical reflections and sensitivity and self-reflections on the purpose of talking to either group. In the DIGINAUTS project, participating researchers deliberately refrained from interviewing in hotspots or camps. One should, however, also have awareness of the fact that even research that nominally challenges bordering practices can inadvertently become big business (see for instance Pendakis 2020 on the massive NGOisation of refugee solidarity in Greece). Many studies have focused on forms of micro-resistance but under-prioritised studying the structural context.

Our point of departure in our research was to use ethnographic and digital material to speak back to system. This brings us to three insights from our studies—here presented as propositions. Firstly that, engaged research is not only about empathising with refugees and migrants and trying to not do harm, or protect them, but also about illuminating the broader social and economic constraints that they face on many scales (not just the biopolitical space where the bordering takes place). Secondly, acknowledging that research does not take place in a vacuum but must also take into account the location of the bordering regime where it takes place. We should acknowledge the differences that exist within the European asylum, migration, and deportation regime. Thirdly, that there is also a need for awareness that academia is turning more and more into an industrialised mass production-based enterprise where churning out papers affords you more cultural capital that you can transform into economic capital, and that many research papers are, in fact, about vulnerable groups (such as asylum seekers, trafficked persons, sex workers, unaccompanied minors, precarious workers, and illegalised migrants). So we must be aware that, because of the fact that our output depends on those experiences, there is a stronger imperative to publish less, better-quality research that is not only descriptive and analytical, but also critical and engaged, and as Stierl suggests, accentuates the crisis rather than fixing it.

Turning our gaze to the particular site that caused the discussion at the conference we mentioned in the introduction, we can first question its purpose, what it is, and what it offers. Secondly, ask if a militant research approach or the AoM approach necessitates the use of online platforms? More importantly—do either legitimise it? The platform was created on Facebook in 2014. All posts are written in Arabic. By April 2020, it had more than 330,000 members. The group is one of several, which all aim to support people who need information and support to navigate the European border regimes (see Fischer and Jørgensen, forthcoming b). Other groups offer information on sea conditions (write and share about the sea, weather conditions) but the platform is by far the biggest of such groups.

Users of the group use it as a tool for sharing information. It seems that people use both their own profiles and profiles made for the occasion to post questions. Often a post is made, a question asked, and soon after answers have been given, the post is deleted. The group is a closed group where one has to apply for membership that is approved or declined by one of the administrators. However, a group with more than 330,000 members is de facto open and it is easy to see how anyone could gain access if they wished, including police and border agencies. This is probably a reason why the group rules stipulate that human smugglers have no access, and neither must deals be made with smugglers or brokers. The platform is both a constantly evolving knowledge base of mobility and an infrastructure of connectivity, which in our understanding is an example of a mobile commons. We chose deliberately to bring this particular mobile commons into our understanding of deportable persons' struggle against deportation and border regime(s). Not to reveal or outline particular strategies but to emphasise and acknowledge how counter-knowledge is produced and utilised to challenge the (also digital) deportation regimes (De Genova and Peutz 2010).

The ethical dilemmas in online ethnographic research are arguably different from on-site ethnographic research. Donath makes an interesting argument here: "In the physical world there is an inherent unity to the self, for the body provides a compelling and convenient definition of identity. The norm is one body, one identity [...]. The virtual world is different. It is composed of information rather than matter" (Donath 1998). This is something that has to be kept in mind while doing online research, namely that it is a two-way street: on the one hand, people might not know that their posts are being used for research; on the other hand, we do not actually know who is sitting at the laptop making these posts, and with what intention, etc.

People posting on the platform have many reasons to use fake identities, in other words rely on deception, to get the information they require for their journeys. Instead of viewing digital technologies exclusively as enabling monitoring and control of migrants, we should also recognise migrants' capacity for deception within the online world and the creative ways in which they gain access to information in the spirit of the autonomy of migration approach. This leads us to another proposition that may be perceived as controversial. We should, if not reject the surveillance bias, then at least be reflexive about it when studying migrants and other vulnerable groups, for it can inadvertently create a condition of permanent victimhood. We should not harvest different closed groups and sites for data or reveal information, but we should communicate that people react and resist and seek to challenge the deportation machinery.

We conclude this chapter by stating that all social science research is in danger of putting others at risk. This is why we need reflexive and contextualised ethics that allow us to conduct the research we believe could criticise and improve a dehumanising system such as the deportation regime. We should be aware not to make research impossible that could assist in highlighting the brutality of such regimes.

BIBLIOGRAPHY

Agustín, Óscar García, and Martin Bak Jørgensen. 2018. *Solidarity and the 'Refugee Crisis' in Europe*. Cham: Springer.

Ali, Suki. 2006. "Racializing Research: Managing Power and Politics?" *Ethnic and Racial Studies* 29 (3): 471–486.

Bell, Claudia. 2017. "'We Feel Like the King and Queen': Western Retirees in Bali, Indonesia." *Asian Journal of Social Science* 45 (3): 271–293.

Birnie, Rutger. 2019. "The Ethics of Resisting Deportation." In *Proceedings of the 2018 ZiF Workshop "Studying Migration Policies at the Interface between Empirical Research and Normative Analysis,"* edited by Matthias Hoesch and Lena Laube, 191–214. ULB Münster (miami.uni-muenster.de). https://doi.org/10.17879/95189423213.

Block, Karen, Elisha Riggs, and Nick Haslam. 2013. *Values and Vulnerabilities: The Ethics of Research with Refugees and Asylum Seekers*. Toowong, Australia: Australian Academic Press

Bojadžijev, Manuela, and Serhat Karakayali. 2010. "Recuperating the Sideshows of Capitalism: The Autonomy of Migration Today". *e-flux journal* 17: 1–9.

Boswell, John, Jack Corbett, Kate Dommett, Will Jennings, Matthew Flinders, R. A. W. Rhodes, and Matthew Wood. 2019. "State of the Field: What Can Political Ethnography Tell Us About Anti-Politics and Democratic Disaffection?" *European Journal of Political Research* 58 (1): 56–71.

Burawoy, Michael. 2005. "For Public Sociology." *American Sociological Review* 70 (1): 4–28.

Casas-Cortes, Maribel, Sebastian Cobarrubias, and John Pickles. 2015. "Riding Routes and Itinerant Borders: Autonomy of Migration and Border Externalization." *Antipode* 47 (4): 894–914.

Clifford, James, and George E. Marcus. 1986. *Writing Culture: The Poetics and Politics of Ethnography*. Berkeley: University of California Press.

De Genova, Nicholas, ed. 2017. *The Borders of "Europe": Autonomy of Migration, Tactics of Bordering*. Durham, NC: Duke University Press.

De Genova, Nicholas, Glenda Garelli, and Martina Tazzioli, eds. 2018. "Autonomy of Asylum? The Autonomy of Migration Undoing the Refugee Crisis Script." *South Atlantic Quarterly* 117 (2): 239–265.

De Genova, Nicholas, and Nathalie Peutz, eds. 2010. *The Deportation Regime Sovereignty, Space, and the Freedom of Movement*. Durham, NC: Duke University Press.

Donath, Judith S. 1998. "Identity and Deception in the Virtual Community." In *Communities in Cyberspace*, edited by Peter Kollock and Marc Smith. London: Routledge.

Düvell, Franck, Anna Triandafyllidou, and Bastian Vollmer. 2010. "Ethical Issues in Irregular Migration Research in Europe." *Population, Space and Place* 16 (3): 227–239.

Fahrendorff, Christian. 2012. "Illegal Indvandring Til Danmark Stiger." Dr.dk, 29.03.2012. Accessed March 11, 2021. http://www.dr.dk/Nyheder/Ind land/2012/03/29/191724.htm.

Fischer, Leandros, and Martin Bak Jørgensen. (Forthcoming a). "Marxist Perspectives on Migration between Autonomy and Hegemony: An Intervention for a Strategic Approach." In *Marxism and Migration*, edited by Genevieve Ritchie, Sara Carpenter, and Shahrzad Mojab.

Fischer, Leandros, and Martin Bak Jørgensen. (Forthcoming b). "Autonomy of Migration in the Age of Deportation—Migrants' Practices against Deportation." In *The Migration Mobile: Border Dissidence, Sociotechnical Resistance and the Construction of Irregularized Migrants*, edited by Vasilis Galis, Martin Bak Jørgensen, and Marie Sandberg. Washington, DC: Rowman & Littlefield.

Flick, Uwe. 2018. *An Introduction to Qualitative Research*. Newbury Park, CA: Sage.

Foley, Douglas. E. 2002. "Critical Ethnography: The Reflexive Turn." *International Journal of Qualitative Studies in Education* 15 (4): 469–490.

Gavanas, Anna, and Inés Calzada. 2016. "Multiplex Migration and Aspects of Precarization: Swedish Retirement Migrants to Spain and Their Service Providers." *Critical Sociology* 42 (7–8): 1003–1016.

Glerup, Marie Rask. 2012. "Ingen kender tal for illegale indvandrere." *Detektor Dr.dk*, May, 2012. http://www.dr.dk/P1/Detektor/Udsendelser/2012/05/01170052.htm.

Hage, Ghassan. 2009. *Waiting Out the Crisis: On Stuckedness and Governmentality*. Carlton, VIC: Melbourne University Press.

Halvorsen, Sam. 2015. "Militant Research Against—and—Beyond Itself: Critical Perspectives from the University and Occupy London." *Area* 47 (4): 466–472.

Jacobsen, Karen, and Loren B. Landau. 2003. "The Dual Imperative in Refugee Research: Some Methodological and Ethical Considerations in Social Science Research on Forced Migration." *Disasters* 27 (3): 185–206.

Juris, Jeffrey. 2007. "Practicing Militant Ethnography with the Movement for Global Resistance in Barcelona." In *Constituent Imagination: Militant Investigations, Collective Theorization*, edited by Stevphen Shukaitis, David Graeber, and Erika Biddle, 11–34. Chico, CA: AK Press.

Juris, Jeffrey. 2014. "Activism: Deviation. Fieldsights—Field Notes." *Cultural Anthropology* Online May 18, 2014.

Jørgensen, Martin Bak. 2012. "Categories of Difference in Science and Policy—Reflections on Academic Practices, Conceptualizations and Knowledge Production." *Qualitative Studies* 3 (2): 78–96.

Jørgensen, Martin Bak. 2019. "'A Goat That Is Already Dead Is No Longer Afraid of Knives': Refugee Mobilizations and Politics of (Necessary) Interference in Hamburg." *Ethnologia Europaea* 49 (1): 41–57.

Jørgensen, Martin Bak, Susi Meret, Annika Lindberg, and José Bayona. 2018. "Reclaiming the Right to Life: Hunger Strikes and Protests in Denmark's Deportation Centres." *Open Democracy*, January 7, 2018. Accessed March 11, 2021. https://www.opendemocracy.net/en/can-europe-make-it/reclaimi/.

Karakayali, Serhat, and Vassilis Tsianos. 2010. "Transnational Migration and the Emergence of the European Border Regime: An Ethnographic Analysis." *European Journal of Social Theory* 13 (3): 373–387.

Krause, Ulrike. 2017. "Researching Forced Migration: Critical Reflections on Research Ethics during Fieldwork." Refugee Studies Centre. Working Paper Series (123).

Kumar, Satendra. 2014. "The Promise of Ethnography for the Study of Politics." *Studies in Indian Politics* 2 (2): 237–242.

Mathers, Andrew, and Mario Novelli. 2007. "Researching Resistance to Neoliberal Globalization: Engaged Ethnography as Solidarity and Praxis." *Globalizations* 4 (2): 229–249.

Mezzadra, Sandro. 2011. "The Gaze of Autonomy: Capitalism, Migration, and Social Struggles." In *The Contested Politics of Mobility: Borderzones and Irregularity*, edited by Vicki Squire, 121–142. London: Routledge.

Mezzadra, Sandro, and Brett Neilson. 2013. *Border as Method, or, the Multiplication of Labor*. Durham, NC, and London: Duke University Press.

Nayak, Anoop. 2006. "After Race: Ethnography, Race and Post-Race Theory." *Ethnic and Racial Studies* 29 (3): 411–430.

Nimführ, Sarah, and Buba Sesay. 2019. "Lost in Limbo? Navigating (Im)mobilities and Practices of Appropriation of Non-deportable Refugees in the Mediterranean Area." *Comparative Migration Studies* 7 (1): 1–19.

Papadopoulos, Dimitris, and Vassilis Tsianos. 2013. "After Citizenship: Autonomy of Migration, Organisational Ontology, and Mobile Commons." *Citizenship Studies* 17 (2): 178–196.

Pendakis, Katherine L. 2020. "Migrant Advocacy under Austerity: Transforming Solidarity in the Greek-Refugee Regime." *Journal of Refugee Studies* online first https://doi.org/10.1093/jrs/fez113.

Pittaway, Eileen, Linda Bartolomei, and Richard Hugman. 2010. "'Stop stealing our stories': The ethics of research with vulnerable groups." *Journal of Human Rights Practice* 2 (2): 229–251.

Polson, Erika. 2016. *Privileged Mobilities: Professional Migration, Geo-Social Media, and a New Global Middle Class*. Hamburg: Peter Lang.

Rodgers, Graeme. 2004. "'Hanging out' with Forced Migrants: Methodological and Ethical Challenges." *Forced Migration Review* 21: 48–49.

Rosenblum, Karen, and Toni-Michelle Travis. 2008. *The Meaning of Difference: American Constructions of Race, Sex and Gender, Social Class, Sexual Orientation, and Disability*. New York, NY: McGraw-Hill Companies.

Rozakou, Katerina. 2019. "'How Did You Get In?' Research Access and Sovereign Power during the 'Migration Crisis' in Greece." *Social Anthropology* 27: 68–83.

Scheper-Hughes, Nancy. 1995. "The Primacy of the Ethical: Propositions for a Militant Anthropology." *Current Anthropology* 36 (3): 409–440.

Shukaitis, S. Stevphen, and David Graeber. 2007. *Constituent Imagination: Militant Investigations/Collective Theorization*. Edinburgh: AK Press.

Siapera, Eugenia, and Creta, Sarah. 2020. "The Ethics of Media Research with Refugees." In *Media Activist Research Ethics*, edited by Sandra Jeppesen, and Paola Sartoretto, 221–248. Cham: Palgrave Macmillan.

Stierl, Maurice. 2018. *Migrant Resistance in Contemporary Europe*. London: Routledge.

Stierl, Maurice. 2020. "Do No Harm? The Impact of Policy on Migration Scholarship." *Environment and Planning C: Politics and Space* online first. https://doi.org/10.1177/2399654420965567.

Team Colors Collective, eds. 2010. *Uses of a Whirlwind: Movement, Movements, and Contemporary Radical Currents in the United States*. Edinburgh: AK Press.

Thomsen, Trine Lund, and Martin Bak Jørgensen. 2012. "Researching Irregular Migration: Concepts, Numbers, and Empirical Findings in a Scandinavian Context." *Nordic Journal of Migration Research* 2 (4): 275–279.

Thomsen, Trine Lund, Martin Bak Jørgensen, Susi Meret, HelleStenum, and Kirsten Hviid, eds. 2010. *Irregular Migration in a Scandinavian Perspective*: Maastricht: Shaker Publishers.

Trimikliniotis, Nicos, Parsanoglou Dimitris, and Vassilis Tsianos. 2016. "Mobile Commons and/in Precarious Spaces: Mapping Migrant Struggles and Social Resistance." *Critical Sociology* 42 (7–8): 1035–1049.

Turton, David. 1996. "Migrants & Refugees: A Mursi Case Study." In *In Search of Cool Ground: War, Flight & Homecoming in Northeast Africa*, edited by Tim Allen, 96–110. London: Africa World Press.

Udlændinge-,Integration- og Boligministeriet. 2016. *Tal på udlændingeområdet pr. 31.08.2016*. København: Udlændinge-, Integration- og Boligministeriet.

Van Houtum, Henk, and Rodrigo Bueno Lacy. 2020. "The Migration Map Trap. On the Invasion Arrows in the Cartography of Migration." *Mobilities* 15 (2): 196–219.

Van Liempt, Ilse, and Veronika Bilger, eds. 2009. *The Ethics of Migration Research Methodology: Dealing with Vulnerable Immigrants*. Sussex, UK: Sussex Academic Press.

Van Liempt, Ilse, and Veronika Bilger. 2018. "Methodological and Ethical Dilemmas in Research Among Smuggled Migrants." In *Qualitative Research in European Migration Studies*, edited by Ricard Zapata-Barrero, and Evren Yalaz, 269–285. Cham: Springer.

Venkatesh, Sudhir Alledi. 2013. "The Reflexive Turn: The Rise of First-Person Ethnography." *The Sociological Quarterly* 54 (1): 3–8.

Wissink, Lieke. 2019. "Material Guides in Ethically Challenging fields: Following Deportation Files." In *Secrecy and Methods in Security Research*, edited by Marieke de Goede, Esmé Bosma, and Polly Pallister-Wilkins, 291–305. London: Routledge.

Yalaz, Evren, and Ricard Zapata-Barrero. 2018. "Mapping the Qualitative Migration Research in Europe: An Exploratory Analysis." In *Qualitative Research in European Migration Studies*, edited by Ricard Zapata-Barrero, and Evren Yalaz, 9–31. Cham: Springer.

Zapata-Barrero, Richard, and Evren Yalaz. 2020. "Qualitative Migration Research Ethics: A Roadmap for Migration Scholars." *Qualitative Research Journal* 20 (3): 269–279.

CHAPTER 7

The Redundant Researcher: Fieldwork, Solidarity, and Migration

Vasilis Galis

INTRODUCTION

"A 24-year-old man from Cameroon was found dead inside the Moria migrant camp on the eastern Aegean island of Lesvos early Tuesday morning [...] as temperatures fell below freezing" (Athens-Macedonian News Agency 2019; *The Guardian* 2019). These were the headlines that I was confronted with four days before my arrival on Chios and Lesvos, the two major islands in the north-east Aegean Sea, at the sea border separating Greece and Turkey, hosting thousands of migrants attempting to reach Europe. I was about to launch my fieldwork, interviewing migrants about their digital practices regarding the European border regime within the framework of the DIGINAUTS project. These were also my first tentative steps concerning research fieldwork that was both new and hostile to me. Migration studies constitutes a rapidly growing field resulting in what Casas-Cortes et al. (2015, 63) call a "migration

V. Galis (✉)
Center for Digital Welfare, IT University
of Copenhagen, Copenhagen, Denmark
e-mail: vgal@itu.dk

© The Author(s) 2022
M. Sandberg et al. (eds.), *Research Methodologies and Ethical Challenges in Digital Migration Studies*, Approaches to Social Inequality and Difference, https://doi.org/10.1007/978-3-030-81226-3_7

knowledge hype". While I was struggling to find my space in the field, without being part of the hype, Greek governmental sources leaked that the causes of death of the young Cameroonian mentioned above were unknown. However, local self-organised solidarity groups, such as the No Border Kitchen, were very clear:

> On January 8, 2019 the European Border regime, and violent neglect of human life and dignity present in Moria camp on the Greek Island of Lesvos led to the death of Jean Paul, a 24-year-old man from Cameroon. The last few weeks have been the coldest ones yet this winter, with temperatures hovering around freezing, high winds, and frequent rainstorms. Much of Moria has been experiencing power outages for days, leaving many people without heat and other basic necessities. (No Border Kitchen 2019)

Already before I left Athens, I had a banal sense of doing something terribly wrong. Who am I, with my White academic privileges, to approach entrapped migrants in the trench of Fortress Europe and talk to them about their use of digital applications? How can I justify, first to myself, politically and even academically, the fact that I was seeking to ask questions about mundane and trivial experiences, while people were literally freezing to death? What is this research about and for whom is it relevant? *This chapter is not an answer. This chapter poses questions.* What is it like to conduct academic research on a phenomenon that is polluted by vested political interests, personal tragedies, ideological loyalties, propaganda, and hazards for the subject of research or on a subject of research in danger? How is this kind of fieldwork compatible with my ideological integrity? Is it possible to do research that contributes to migrant struggles? Am I performing action research, or just building my career?

The aim of this chapter is to reflect on and investigate the possibility of conducting migration research that contributes to the freedom of movement and safety of mobile populations, while reflecting on the emotions and political loyalties of the researcher (see also Fischer and Bak Jørgensen, Chapter 6 in this volume). This chapter also aims to contribute to the epistemological/methodological debate on migration studies by suggesting concrete principles for an *emancipatory migration research paradigm*, building on the combination of disability studies and the Autonomy of Migration (AoM) approach (for an introduction to the AoM approach, see Chapter 6). To accomplish that, I will first relate personal experiences and emotions, reflections, and narrations by migrants and solidarians[1] during and after qualitative fieldwork on the Greek islands of Lesvos and Chios in the winter of 2019. These experiences, emotions, reflections, and narrations are "constricted by the historical context in which they are made" (Nordstrom and Robben 1995). This period was almost four years after the 2015 so-called "long summer of migration" when a large migratory influx reached the European borders, with mainly newcomers from Syria. Thousands of migrants were entrapped (and still are) in open and closed detention centres or hotspots[2] in the Greek islands, the EU-Turkey Statement[3] was in force, and a wave of anti-migration voices and false information spreading through social media and other digital sources had taken over the public debate both in Greece and internationally (Farkas et al. 2018; D'Haenens et al. 2019; Titley 2019). In the same period, both the Greek state and mainstream media launched a whisper campaign against solidarity with migrants. NGOs and self-organised collectives were criticised, accused, and discredited by government officials and reporters/journalists who sympathised with the regime (Fekete 2018; Rozakou 2017, 2018; Gordon and Larsen 2020).

An extended network of solidarity, comprising mainly political actors from the extra-parliamentary spectrum, was initiated to welcome the newcomers, confront racist and xenophobic behaviours, and accommodate migrants during their temporary or permanent stays in Greece. These activists and solidarians voiced scepticism to researchers and academics visiting the islands to conduct research on the recent flows of migrants. Questions arose about who would benefit from this research, in what ways the mobile subjects would be supported by the academic research, what constitutes action research, and whether there is a risk that we all fall into the trap of action research washing. Given that a good part of migration research supports policy (Black 2001) and that my research agenda *did not* aim to provide any policy recommendations, doing fieldwork resembled a political, ideological, methodological, and epistemological minefield, with every step signalling a warning in all kinds of directions. I was about to conduct research on vulnerable migratory subjects exposed to physical and structural violence, issues of trust and power asymmetries between the researcher and the researched were prominent, and my own ideological/political concerns as well as emotions were kicking in. What a mess! In the following, I outline my research stays at the islands of Chios and Lesvos. The description of the fieldwork will be enhanced by reflections and concerns that emerged while conducting the research. I then translate these concerns into a concrete set of principles for conducting emancipatory migration research, inspired by disability studies. The chapter will conclude with a final reflection on conducting politically engaged research.

Arriving on the Islands: From Emotions of Scepticism to Holistic Shame

Despite my long-lasting aerophobia, I took the early flight from Athens to Chios on 12 January, which took almost 30 minutes. I soon realised that I was making the same journey as many migrants dreamed of making but in reverse and in the most convenient and less time-consuming way, instead of spending hours on a boat in the middle of winter. I was travelling from the Greek mainland, a European capitol, and one of the first urban stops for migrants, to the north-eastern islands on the border with Turkey. There I was, with my European mobility and class privileges, travelling safely, fast, and comfortably to one of the epicentres of modern migratory drama, where thousands of people were literally trapped in camps, detention centres, and hotspots under horrific conditions in terms of hygiene, weather, and freedom of mobility. On reaching Chios, I was welcomed by an old colleague of mine who showed me around and made sure I would experience the finest of the local cuisine. The emotional roller coaster was about to begin—and was not to be underestimated. Even in the literature on reflexive research practice, emotions tend to be overlooked (Gray 2008). Therefore, I turn to feminist and queer epistemologies that have problematised researchers' own reflections on dealing with emotions, not only as an impact on the researched or the research agenda, but also on themselves (Van Liempt and Bilger 2009). Following Ahmed (2014, 4), dealing with emotions in research practice and about the research subject "is clearly dependent on relations of power, which endow 'others' with meaning and value". My privileged position and ability to move on Chios seriously impacted my feelings for my fieldwork and ascribed vulnerability to the migrants. My subject position was co-constructed by my emotions and scepticism regarding my research but also through the experience of vulnerable populations detained in a huge hotspot surrounded by water. I will return to the concept of vulnerability later.

My first hours on Chios were impregnated by strong scepticism and a feeling of personal *shame*, reflecting on the asymmetries between my subject position and the migrants whom I was about to meet. What on earth was I doing here? How would I be able to find and talk with migrants living in inhuman conditions about mundane digital practices while I was enjoying my host's hospitality, my freedom of mobility, and the warmth of my hotel room? How could I convey that to my political circles? Is it possible to separate research (work) from political standpoints and my own situatedness? It was impossible for me to conceal or separate my emotions from my way of thinking or even the way I was conducting my fieldwork. Gray (2008) argues that academic inquiry is concurrently an embodied, emotional, and political activity, and therefore emotions also partake in knowledge production. In that sense, my emotions also reflected my standpoint on the world, or how I am apprehending migration issues in general, not merely as a researcher or a politically active subject, but as a whole. Through my emotions, I reacted to the contradictions I faced upon arriving on the islands. This was not a martyr's act of self-flagellation, but rather a realisation of how "I" and "we, the research community" are shaped by our contact with the ontologies of migrants. In that sense, these kinds of emotions also involve politics, since they constitute reactions to how power relations enact our fieldwork ontologies and bring epistemological attention to how we, as researchers, become invested in specific issues (cf. Ahmed 2014). But is this enough? Will sharing these reflections and emotions in another academic paper accessible to a specific readership and using sophisticated literature make any difference, and to whom?

My first couple of days on Chios were spent on developing a snowball effect to find relevant informants for the study. The target groups were migrants and solidarians who were keen on using digital media or hosted websites, social media accounts, or other self-organised media, which facilitated solidarity with mobile populations and produced practical information. To do that, I employed my local contacts and acquaintances, especially through my colleague who lived on the island, and my efforts were quite successful. For reasons of ethical and political integrity, I consciously avoided contacting people whom I knew through political activity related to solidarity work with migrants, but soon realised that separating worlds would be impossible. Although I wanted to avoid mixing up these categories, the reality of the fieldwork brought together myself (as a researcher), solidarians whom I perceive as political comrades, migrants involved in solidarity projects, and therefore politicised, and myself (as a conscious political subject). How could it be possible otherwise? My subject positions constantly changed, oscillating between an anticipated academic distance that many epistemological traditions require (neutrality I believe it is called) and my political engagement (partiality).[4]

What I felt was shame, a feeling of corrosive shame and betrayal for performing a research role in a heavily politicised field with which I usually practiced solidarity. My reflections on university careers, elitist academic research detached from the actual needs of migrants on the move, redundant research questions or problems concerning the hardship of living detained in tents in the middle of winter, and many other critical thoughts dominated my thoughts. Ahmed (2014) argues that shame prevents the individual from betraying ideals, while the lived emotion of shame makes the individual understand and appreciate the reasons for adopting these ideals in the first place. Feeling shame is a manifestation of failing to achieve these social ideals but it also allows us to reflect on and come closer to what we are failing to accomplish. According to Ahmed, the feeling of shame can be restorative if it is temporary because "shame binds us to others in how we are affected by our failure to 'live up' to those others, a failure that must be witnessed, as well as being seen as temporary, in order to allow us to re-enter the family or community" (ibid., 107). I struggled and kept struggling to find the restorative element in feeling shame during my fieldwork and the ways to bridge research with my political situatedness. Is this even possible? I will return to that.

FIELDWORK OR MINEFIELD? RELUCTANT TRUST, VULNERABILISATION, AND POLITICAL SOLIDARITY

When the interviews began, I faced the raw reality of migrants being entrapped on an island in the middle of winter. I talked mainly to young males with different backgrounds, ethnicities, and sexual orientation. These interviews were emotionally and empirically strong, touching upon mundane aspects of everyday life in the detention centres, the hotspots, and on the island in general through the lens of digital media usage. It was obvious that the use of smartphones and social media played a vital role in the well-being of migrants as well as posing a threat for a variety of reasons, which are not part of the scope of this chapter. However, this meant that conducting this particular fieldwork was not as redundant as I had thought from the outset. My informants were very eager and open about discussing their digital habits and found it amusing to show me different applications as well as tricks they employed to gain internet access. This made me immediately concerned. Suddenly, I had access to sensitive information about closed social media groups, methods to safely cross the borders, secure chats, tricks on how to use the smartphone, ways to acquire SIM cards, and so on. This was not a problem in terms of the privacy of the migrants themselves or their protection. I was perfectly aware of the sensitivity of the material and ways to protect it or omit it from my empirical storage. I turned off the recorder when sensitive information was communicated and/or I excluded from the transcription or the analysis knowledge that could harm my informants.[5] What worried me and made me reflective was the ease and trust that these people showed me. Usually, simply the word "research" might create mistrust and raise suspicion among informants.

The common issue in research situations like this is how to build trust and to invest time in establishing personal contacts with possible respondents (Van Liempt and Bilger 2018). I experienced exactly the opposite phenomenon. Due to my personal acquaintances and my political involvement (which was not consciously activated to create the snowball effect, but did make a difference after all), I was granted access to a large group of migrants (and solidarians) who were eager to openly talk to me about their personal encounters with digital media, describing in detail important and sensitive information. The feeling of shame and bad conscious dramatically returned. What do I give back to my research informants? Being conscious of the power asymmetries, me with my academic and European privileges collecting material on their digital habits and furthering my research agenda, and migrants in their vulnerable positions entrusting me with their information capital, for what?

Things became even more complicated when some of the informants wanted to be friends with me on social media, which I did not decline. Within a few hours of the interviews, I received several messages with various content: from friendly discussion to asking for help and money to legal advice. It was obvious that my informants perceived me as a person who could help them improve their position due to my intersectional subject position (see also Düvell et al. 2010). Right there, the power asymmetries were visualised and concretised in various texts in Messenger. Several questions arose: how did the way they perceived me influence what they told me? How should I assess the quality of information, given that the subject is in a vulnerable position? What should my role be, not as a fragmented (sometimes a researcher, sometimes an activist, sometimes neither) subject but as a whole, to their requests whatever these were? Was it wrong of me to digitally connect with these people? Wrong for whom and for what? Was it politically, socially, or research-ethically wrong? According to Van Liempt and Bilger (2018, 278) "researchers in this field must be aware that the relation between the researcher and the respondent, even if trustful and close, is not equal and is clearly influenced by inequalities of rights, legal and economic position, gender and/or psychological position". What if the researcher refuses her one-dimensional role and reflects on her pluralist subject? What happens then to the quality of research or the political activity? These are issues that, along with feminist theory, anthropology, and more (e.g. Clifford and Marcus 1986; Foley 2002; Venkatesh 2013), I will problematise in the next two sections. Ahmed (2014) again explains that the swarm of emotions partaking in the research process shape interaction with informants and create boundaries for the research subject. These emotions "produce the very surfaces and boundaries that allow the individual and the social to be delineated as if they are objects" (ibid., 10). In that way, my pluralist subject position also populates the epistemological arena of my research and colours my interpretation of the social in terms of migration and political solidarity.

What made me additionally concerned was the fact that I constantly experienced my informants as in a vulnerable position. Staying true to the idea and practice of solidarity,[6] the "vulnerabilisation" of migrants, either by my own feelings or by their given situation, gave me a sense of unease. According to Lind (2020, 45), vulnerabilisation comprises "the processes (in the context of migration and beyond) of constructing, attributing and governing vulnerability". Did I (a White, middle-class, heterosexual cisgender man), or my research, constitute my informants as vulnerable? While Lind refers to how specific migratory groups should be attended to, cared for, or governed, I reflected on my own concern regarding how the contradiction between humanitarian empathy and collective solidarity works to oppose the structural racism and violence that moving populations face upon arrival at Fortress Europe. In other words, conducting fieldwork in this case activated a feeling of superfluousness and powerlessness on my behalf that automatically turned my informants into vulnerable subjects worth of empathy. In addition, some of the interviews were strong enough to trigger distressing memories among the informants. I, a western researcher, collected stories and narratives about forced migration, war, abuse, loss and grief, survival, and forms of violence that non-western people had experienced. I found myself in a situation where my informants were affected by recalling and reproducing painful dimensions of their lives and I oscillated between emotions of (humanitarian) empathy and my values of political solidarity.

By this I mean that in this context, I perceive solidarity as a horizontal coalition of migrants, activists, and local people, leading to a series of actions that place migrants' desires and passages at the heart of the action, instead of framing migrants in humanitarian terms of being victims and vulnerable. I could not find myself in the humanitarian responses to global oppressions through gestures of compassionate hospitality (see also Tyler 2006; Millner 2011) or the human rights discourse. In the western liberal narrative, someone who is worth rights, empathy, help, charity is someone innocent, good, exposed to harm. There is an element of purity in the vulnerable subject (see also Ticktin 2017). It is almost a Christian theological thesis that impregnates the western human rights discourse. Lind (2020), following Arendt (1951), explains that human rights rely on citizenship and on belonging to a community that attributes rights. This is a political and not a humanitarian issue. Turning to human rights and a humanitarian approach constitutes an act of depoliticisation (Ticktin 2014) through delegating solidarity with migrants to state and institutional charities. Charity is the humanitarian mask behind the face of economic exploitation (Žižek 2008, 19). This also implies recognition of formal systems of classification that divide moving populations into economic migrants, refugees, asylum seekers, deportable, legal and illegal individuals, and so on (see also Jørgensen 2012; Lundberg and Söderman 2015).

In the context of the liberal state, the human rights discourse, humanitarianism, and philanthropy maintain the power hierarchies between those who have rights and those who claim rights, and the administrative and statistical distinction between those included and those excluded from the privileges and rights provided by the nation-state. Vandevoordt (2019) explains that, as human rights providers or beneficiaries are perceived to acquire resources, power, and expertise, human rights receivers are reduced to vulnerable individuals in need of being fed, cared for, and represented by others. In the context of my fieldwork, the increasing criminalisation of self-organised solidarity work by the Greek authorities and the "NGOification of solidarity", meaning the normalisation and assimilation of grassroots solidarity by professional and institutionalised international humanitarian organisations, added another element of antagonism between humanitarianism and grassroots solidarity (Rozakou 2017).

My understanding is that engaging with migration issues implies a politics that sides with minorities, the stateless, the powerless, the undocumented, the "undesirable" migrants, those who are not protected by the constructed state borders and territorial lines (see also Honig 2009; Millner 2011; Lind 2020). The vulnerabilisation of my informants, through the research process, belonged to a discourse and practice that were alien to me. One of the pertinent questions that arose for me during my fieldwork on Chios was how political solidarity, in the terms described above, can be compatible with academic research that cannot only be considered redundant but also may create emotions of shame as well as "vulnerabilising" the research subject. Düvell et al. (2010) argue that (migration) researchers are primarily not activists but academics and they are responsible for high-quality and ethical research, their subjects, their profession, their funders, and so on. For the readers who side with the latter and adopt a symmetrical or neutral epistemological research standpoint, I would respond that all research is partisan. This constitutes a step beyond the dualist epistemological dilemma of modernism, that is, being epistemologically objective about a responsible and value-free science versus being epistemologically objective about an ethically engaged science (Galis and Hansson 2012). The issue here is what kinds of emotions and subject positions the research process enacts and how these partake in the constitution of the researcher's subjectivity and/or how they become (in)compatible with what is called solidarity in radical political terms or "shared conversations" in epistemology (cf. Haraway 2001, 176). Therefore, and given the partiality of all actors involved in migration research, the challenge for me here was to substitute or convert the emotion of shame into epistemological and political solidarity with the migrants.

I left Chios with these paradoxical reflections, taking the night boat to Lesvos. It was a rainy night and the first impression I had while waiting to embark from the boat at the port of Chios was of a few military vehicles disembarking the huge vessel that had just arrived from Piraeus. This image of an interwar period immediately reminded me that I was not only situated, politically and geographically, at the heart of a contemporary migratory drama but also at the borderland that signifies the diachronic low-intensity cold war taking place between Greece and Turkey. This was also the scene met by thousands of migrants who had left their countries to escape the adversities of war, violence, and poverty. They were not only confronted with the structural violence embedded in bordering practices and migratory policies, but also the visual and mundane violence projected on military technologies and troops guarding the Greek sovereignty. In line with Nordstrom and Robben (1995), the most pressing reality of ethnographically studying populations exposed to violence is the sociopolitical violence prominent where civilian populations are located, such as in camps and hotspots, and in the sociotechnical processes, sceneries, and mundane practices where these populations live and to which they are exposed. My fieldwork also covered violence.

Lesvos: "Hell on Earth"

More than 12,000 people – mainly from Syria, Afghanistan, and Iraq – live in Moria camp, which has grown to become the island's second largest town in just three years. The woman's death on Sunday was the third there in two months. An Afghan teenager was killed in a fight in August and a five-year-old Afghan boy was accidentally run over by a truck while playing in a cardboard box outside the camp in September. Holding signs reading 'Moria is hell' and 'We want security and freedom', the protesters were prevented from marching farther than a few hundred metres (yards) from the camp's gates by around two dozen riot police. Moria, in a former military base, opened in 2015 as a centre to register new arrivals but is now at four times its capacity and it has spilled over into a muddy, garbage-strewn olive grove. (Reuters 2019)

The excerpt above from an international mainstream corporate media, such as Reuters, is indicative of the violence that migrants are exposed to in the camps established by the Greek government with the financial support of the EU. Moria, on Lesvos, was one of Greece's largest migrant camps. There are several others in the Greek islands and mainland. Arriving on Lesvos made me realise that the whole island resembled a huge camp or rather a prison for migrants. Migrants living in extremely poor conditions, people dying due to extreme weather conditions, ethnic violence among migrant groups, sexist violence against sexual minorities, racist violence against migrants by local nationalists, structural violence against the newcomers by national authorities, and the European border management apparatus constituted my fieldwork. Moria, and Lesvos in general, is not a hotspot, a migration camp, a reception centre. I could not stop thinking of Andrea Pitzer's (2017) description of camps: "camps have been in existence continuously somewhere on the globe for more than a hundred years. Barracks and barbed wire remain their most familiar symbols, but a camp is defined more by its detainees than by any physical feature. *A concentration camp exists wherever a government holds groups of civilians outside the normal legal process – sometimes to segregate people considered foreigners or outsiders, sometimes to punish*" (ibid., 5, emphasis mine). The organisation of social and mundane life on the islands segregates the local population from newcomers, who are not free to move around or leave the islands.

As on Chios, I spent the first couple of days on Lesvos initiating a snowball effect to identify relevant informants. I talked to young male migrants from Palestine, Iraq, and Cameroon as well as solidarians from Spain and Greece. Most of the migrants I talked to had left their countries of origin because of war and poverty as well as exposure to sexist violence because of their sexual orientation. Even their descriptions of their use of smartphones and social media included an element of violence. Some of them avoided dating applications in their home countries or along the journey for fear of surveillance and suppression. Certain forms of sexuality were illegal in their countries or in transit. Others described how they were receiving threats in digital forums from co-patriots or members of other ethnic groups while on Lesvos. A few of them even described incidents of abuse on Lesvos or the violence in the Moria camp (including a minor insurrection that led to part of the camp being burned down[7]) that they were afraid to post about on their social media in case the authorities found out. Taking pictures or videos inside the camp and spreading them was banned. That would obviously harm the humanitarian face of both the Greek state and the EU. The most shocking descriptions, however, were those concerning the use of smartphones and social media while being at sea, crossing the Aegean to reach Lesvos in small rubber boats that often sank. These dramatic narratives referred to moments of desperation and survival while migrants tried to maintain online contact with solidarians at the other end of the phone who were navigating the trip at sea and whom they could call for help if needed. I was told about parents who lost their children, brothers who lost their sisters, children who lost their mothers in the huge graveyard of the Aegean Sea.

Again, I found my research role rather redundant in this setting. Anthropologist Gayatri Spivak (1988) questions the motives and sincerity of western researchers in studying non-westerners exposed to power asymmetries and violence unless they go through a self-critical questioning not only of their research role but also their cultural subjectivity as historical products of specific privileges (see also Nordstrom and Robben 1995). Is this the purpose of this text? What does reflection mean in the context of the violence and incarceration that my informants are exposed to and how does it make my research sincere and motivated? How many of my informants will actually read these lines and how will this text make a difference to their everyday life? No matter our academic dedication, even to what we have coined as action research, we cannot avoid the legacy of our privileged position and hegemonic culture (cf. Nordstrom and Robben 1995). The structural and material violence that migrants are subjected to is not simply a case of western arrogance or cruelty. They are exposed to the obscene underside of our privileged western culture, which at the same time acts as the necessary supplement to a liberal understanding of dignity, freedom, mobility, and citizenship. The exception proves the rule. And what is the semiotic load of scientific inquiry in this? Does research provide any social change for migrants or does it perpetuate, in a critical way, of course, the very core of the European border regime that sustains Fortress Europe?

In the final days of my stay on Lesvos, I conducted interviews with a couple of solidarians. Their descriptions were vivid and sober, slightly distant from the experiential narratives of the migrants. However, the element of violence was also dominant. They provided me with short and contemporary historiographic accounts of the solidarity movement on the island and how it diachronically dealt with the systemic violence inherent in the border regime. The solidarians also described the agony and traumatic experiences that migrants experienced in captivity or upon arrival or during their stay in the island. It was obvious that it was not the first time they had talked to researchers and journalists. My political background motivated them to talk to me, despite their disillusioned approach to academic research and mainstream journalism. They had no expectations or hope that once more telling the story of how Lesvos has turned into a living hell for migrants would make any difference. This was common knowledge by then, and several scholars and journalists (see, for example, the *New York Times* 2018; Al Jazeera 2018; Deutsche Welle 2019; *The Guardian* 2020; Balouziyeh 2017; Colson 2017) had written about the issue without any profound changes or interventions to improve everyday life for the migrants. This made me think of Hannah Arendt's (1972, 132) words: "the ceaseless, senseless demand for original scholarship in a number of fields, where only erudition is now possible, has led either to sheer irrelevancy, the famous knowing of more and more about less and less, or to the development of a pseudo-scholarship which actually destroys its object". The aim of the DIGINAUTS project that I was part of was to highlight how migrants' widespread, varied, and innovative digital practices remake migration and potentially create networks of solidarity as migrants navigate through the European border regime. How relevant was this issue and to whom? If social scientific fieldwork is characterised by a combination of empathy and detachment (Robben 1995), who would benefit from conducting scholarship in this field, and what impact would I have on the subject/object of the research?

I have two choices here. *Either* to sink into this bottomless academic/existential fatalism that scorns my research and academic research in general. Badiou (2004) provocatively states that "it is better to do nothing than to contribute to the invention of formal ways of rendering visible that which Empire already recognizes as existent". Žižek (2008), building on Badiou, goes one step further, discouraging scholars from engaging with debates and research that allow the system to run more smoothly. *Or* to choose not to stay inactive and to move the research agenda beyond a vulnerabilisation framework and an internal academic self-confirming that builds professional careers and enhances the intellectual debate with "fascinating" empirical material and flashy theoretical concepts. After all, "researching other people's sufferings can only be justified if the research explicitly aims at alleviating that suffering" (Turton 1996, 96). Lind (2020), in describing his epistemological and methodological standpoint while writing his doctoral thesis, argues for research that can contribute reciprocal benefits that include its participants, such as "taking responsibility as an activist researcher for presenting analyses that highlight injustices and hopefully make social change in a more democratic and inclusive direction possible" (ibid., 83). Pittaway et al. (2010) propose a method of working with vulnerable populations in an anti-oppressive way, despite the subject position imbalances. Van Liempt and Bilger (2018) call for subject-centric methodological approaches in research fields, such as those of migration studies. For me, this implies developing a research agenda that takes responsibility for social interests in the production of knowledge and deconstructs powerful actors such as the state, the law, the border, and orthodox approaches to research.

This agenda must openly resonate with the research subject and how the research subject ought to participate in the configuration of the research work and its implications. We need to return to our informants, in whatever form possible, the information and cognitive capital that they generously offer us. In the context of migration research and border studies, I do not suggest another policy discussion or recommendation concerning the democratisation of institutional processes and the border regime. Along with Lundberg and Strange (2017), I aim towards a "post-institutionalist" take on migration politics and human rights that departs from the importance of everyday acts in providing a political grounding for migration-related research. In that sense, I believe that borders are vectors of specific politics that cannot be subject to democratisation. I argue in favour of methods and a research epistemology that reconstitute social relations between the researcher and the researched, the subject and the object.

Inspired by the Autonomy of Migration (AoM) approach, I want to contribute a methodological framework "that prioritizes the subjective practices, the desires, the expectations, and the behaviors of migrants themselves" (Mezzadra 2011, 121), not as in a romantic and idealist approach to migration studies methodologies, which often implies empty buzzwords that scholars tend to use in ambitious and action-oriented research projects. In line with Fujimura (1991, 223), I "want to take stands, to take points... I want to construct concepts and theories to help some people win over others". I want to implicate my subjectivity and my methods in the epistemological realm of my research and therefore partic-ipate in the presentation, management, and politicisation of the topic under investigation (see also Mackenzie 2012). This does not constitute a methodological invention by any means. Several fields have reflectively recognised the participation of research subjectivities in research (Marres 2012). My aspiration here is to go beyond the recognition of my own situatedness. I am interested in systematising the ways that method can actively engage with the research outcome. I want to explore the potential of methods to "contribute to the framing of change" (Lury and Wake-ford 2012). The question then becomes, how do our methods intervene, interfere, and/or refract in the knowledge-producing debates of which they are part?

BACK TO ATHENS: TOWARDS AN EMANCIPATORY MIGRATION RESEARCH PARADIGM?

I took the late afternoon flight from Lesvos to Athens. My forehead was hot, and I felt dizzy, not only because of the epistemological and methodological riddle overwhelming my mind. I had caught a serious cold that reminded me that soon enough I would return to the comfort and ease of my apartment in Athens in contrast to most of my informants, who would continue dwelling in the cold tents in the Moria camp or wherever they were located. I would spend the next couple of days in bed trying to recover from a regular flu, which gave me plenty of time to reflect and organise my thoughts around the issues thoroughly described above. This translated into a concrete idea to contribute to the valuable AoM approach with a concrete vocabulary and set of methodological actions that may assist the struggle of migrants for emancipation, free mobility, and open borders. Casas-Cortes et al. (2015) explain that politically engaged investigation in migration studies has two main tasks. Firstly, to identify and analytically and politically support contested politics with which migrants engage, the conflicts and ruptures that migration practices cause, migrants' strategies to enable all kinds of movement, and the migrants' experiences. Secondly, to reshuffle the epistemological standards of migration research methodology by turning migrants into subjects rather objects of research, management, care, advocacy, and so on and by simultaneously removing the status of migration researchers as advocates who speak for, activist scholars and scholar activists who act on behalf of others. Accordingly, one of the main demands I made of myself was to refrain from developing an epistemological and methodological toolkit stemming from my own subject position, the western privileged white, cis-researcher. How could I avoid the pitfall of becoming an avant-garde translator/interpreter of migrants' needs, ideas, and will in the research sphere? For many years, I have worked and researched within disability studies. At the start of the 1990s, disability scholar Mike Oliver coined the term *emancipatory disability research* to introduce a radical new approach to researching disability. Oliver (1992), a disabled person himself, suggested that disability researchers must interact with disabled people and their organisations on a regular basis and must enlist their knowledge and skills at the disposal of disabled people for them to use in whatever ways they choose. Could this be translated into migration research? This will be a modest attempt.

There is a long tradition of participatory research methodologies within disability studies. However, these kinds of methodologies have a normalising nature. For instance, the growth of participatory approaches with people who have learning difficulties has been dominated by the rhetoric of normalisation and community care (French and Swain 1997). Similarly, migration studies has been dominated by a normalising governance discourse, enforced by an allegedly politically neutral methodological format through migration narratives, policy mobility frameworks, and technical contributions. In that way, "research protocols in migration studies are standardized and reconstituted as objects of disciplinary investigation and the political and social stakes involved in migrant advocacy are 'professionalized' and diluted" (Casas-Corteset et al. 2015, 63). This is not my intention. I oppose the normalising discourses of the integration of migrants in the western labour market and/or culture as well as technologies and policies of migration management and control. I am sceptical of standardised research protocols that perform politically "neutral" inquiries and contribute sound policy recommendations.

The emancipatory disability researcher engages in political action by changing the relationships involved in research production, i.e. the power relationship between researchers and the researched. Emancipatory research, within disability studies, stems from the politics of the disability movement and aligns with the social model of disability, which perceives disability not as a medical entity or an individual problem, but as a social and political issue that permeates disabling infrastructures and cultures. One can argue that emancipatory disability research is not a research methodology as such, but rather another political instrument at the disposal of disabled people in their struggle to control decision-making processes that shape their lives, to counteract societal and cultural biases, to intervene in the built environment, and co-produce accessible infrastructures (see also Galis 2006). In that sense, I see similarities with the conceptual framework of the AoM approach that views migration not primarily as a phenomenon defined by state power or the discourse on sovereignty but rather as a political and social movement itself (De Genova 2017). An emancipatory methodological toolkit in line with AoM should then enable and encourage migrants to configure the research agenda.

Specifically, a major characteristic of emancipatory research is the insistence that migrants should control (rather than merely participate in) the entire research process from the formulation of the research question to the dissemination of the findings. Researchers are thus at the service and under the direction of people on the move who are no longer "the researched" but rather co-researchers and managers of the research. Emancipatory research adopts the AoM approach whereby the focus for research is that "migration constitutes an essential field of research that allows us to critically understand capitalism. There is no capitalism without migration, one could say, with the regime that attempts to control or tame the mobility of labor playing a strategic role in the constitution of capitalism and class relations" (Mezzadra 2011, 125). With that said, an emancipatory migration research paradigm is meant to be part of the struggle against state and capitalist repression and exploitation and considers borders scars on the body of Earth. Inspired by disability studies and the emancipatory research paradigm (cf. French and Swain 1997), I believe that critical and politically engaged research on migration needs to address and be evaluated against a number of principles that can be stated as questions:

1. Does the research promote migrants' control or scrutiny over the processes that shape their lives?
2. Does the research address the concerns of migrants themselves?
3. Does the research support migrants in their struggle against oppression and for free mobility?
4. Does the research guarantee the safety and integrity of migrants?

Research on migration is unlikely to undergo any substantial change without more fundamental changes in the way migration and migrants are viewed within academia. Moving populations are being empowered by themselves and the political solidarity networks that support them. The question is, can migration researchers become part of that empowerment? This can raise significant critique within the academic community regarding the issue of political neutrality versus academic partisanship. Even for social constructivists, knowledge production must be protected by commitment to political values via the researcher's adoption of symmetrical and neutral stance. Instead, I view the research practice as part of complex networks of practices and struggles subject to power and economic relationships (May 1994). The researcher may be drawn in as a participant or used as a tool (Scott et al. 1990). Researchers have never been isolated from politics, as they have always engaged with diverse groups when conducting research and have returned their findings to these groups (Burawoy 2004). I advocate an emancipatory migration research paradigm that allies itself with struggles and solidarity movements related to migration and moving populations. Disability activist and academic Colin Barnes (2003), in a reflective paper about the impact of emancipatory disability research, argues that "research outcomes in themselves will not bring about meaningful political and social transformation, but they must reinforce and help stimulate further the demand for change. Hence, the main targets for emancipatory disability research are disabled people and their allies" (ibid., 13). Without involving migrants, their allies, and their political targets in this epistemological pursuit, we will merely reproduce academic hierarchies with researchers in the lead. In other words, when directly linked with migrants' ongoing struggle for free mobility, decent living conditions, open borders, papers for all, conducting emancipatory migration research can have a meaningful impact on the empowerment and policies affecting migrants' lives.

I see three lines of criticism here: (1) Can this be done? How can we involve migrants in configuring research agendas and how can we encourage the institutions that fund our research to support this idea? We need to acknowledge and learn from earlier examples of engaged or collaborative research (see for example Sillitoe 2016). The rich tradition and heritage of engaged scholarship must be translated into concrete methodological tools. Thus, it is not only an ontological issue about the empirics of our research but also about epistemological politics and how we as researchers challenge and change the criteria and borders of the political economy of scientific research by establishing practices, political agendas, and actions that create space for emancipatory research within our institutions. (2) Is this an overly Eurocentric approach? Indeed, the experiences and literature presented in this chapter are somewhat "European". I do not argue that this idea necessarily has a universal application. It is useful and compatible with such an approach to compare and juxtapose it to non-western narratives, epistemological politics, and methodological approaches related to migration research. It is more than imperative to queer our western privileged cultural and research capital in the context of doing emancipatory migration research. This is again an onto-epistemological game. As Sandro Mezzadra (2011) claims: "This is helpful not only in itself, but also in order to problematize the way in which we analyze migration in Europe and the 'west'; in order that we methodologically train and decenter our critical gaze" (122). (3) However, this would appear to be indirect recognition of scientific research as the only legitimate game in town? Are there any other ways to empower moving populations and conduct investigative work on migration? I do not recognise academic research as the only way to discursively and analytically support the struggles of migrants. We have a duty as political subjects and researchers not only to engage migrants in the research process but also to co-develop a new political epistemology of migrations that implies methods and ways to return and diffuse this knowledge to the migrants themselves.

Radical seminars and workshops in locations friendly to migrants, circulation through open media and popularisation of the knowledge acquired, online networks and discussion platforms, activists' meetings, websites to circulate counter-knowledge, and collective discussions (e.g. storiemigranti.org, bordermonitoring.eu, watchthemed.net, kritnet.org, migreurop.org) (see Casas-Cortes et al. 2015) are essential elements of an emancipatory migration research paradigm as well as the ways and suggestions proposed by migrants. This implies a mixed-method approach that not only includes scientific methodologies, such as documentation of experiences and barriers, monitoring and barometering of migrant grassroot struggles, alter-visualisation of counter-mapping, and the production of new concepts (ibid.), but also requires migrant-friendly practices, such as self-organisation and protection of moving populations, especially those without papers or those exposed to violence.

CONCLUSION?

This chapter did not intend to reinvent the wheel of methodological reflexivity for the role of the researcher. It constitutes an uproar built on personal reflections and emotions while doing fieldwork research for the DIGINAUTS project. These are common thoughts noted by numerous researchers in the field and several scholars have problematised them in a myriad of far more sophisticated ways than mine. This is my humble and perhaps uncensored effort to position myself in these debates. However, I failed in one of the major principles of writing an academic paper, that is, to pose as many questions as I can answer within the margins of the manuscript. But as I stated in the introduction, this chapter does not necessarily provide many answers. It certainly poses multiple questions that might be useful to colleagues, especially younger ones and those who are new to the hardship of conducting social scientific research concerning somewhat existential and political aspects of the research process. I questioned my role in terms of intersectional subjectivity and conducting research on migrants living in horrendous conditions. I asked about the compatibility of conducting (symmetrical/neutral) research while being a political subject in solidarity with the research subject. I questioned the triviality, the vulgarity, and relevance of the research topic considering the brutality of the living status of the "researched". I reflected upon the complexity of the topic and the values, interests, and politics embedded in the fieldwork. I pondered the efficiency of the research for the lives and well-being of migrants and the limits between action research and creating an academic career. I even questioned the point of this paper.

It was important to me to implicate, in a detailed and meticulous way, my own feelings in the description of the fieldwork, to add another blow to the picture of the researcher as an emotionless, neutral, symmetric creature. This fieldwork was mostly about emotions, my own and the migrants' and solidarians' with whom I met and interacted. I extensively and repeatedly referred to my emotion of shame and my researcher's role as redundant. I explained how my political loyalties clashed with my research activity, awakening a deep feeling of embarrassment for conducting what I perceived then as a research investigation that was meaningless for the migrants but valuable to my career. But as feminist and queer scholarship has shown that emotions play an important role in politics (Ahmed 2014), my feeling of shame also acted as a reminder of not only how power asymmetries in fieldwork shape the research, but also that emotions can trigger a conscious change in the epistemological balance of the research process. With this elevating reflection in mind, I turned to my previous experience with disability studies to gain inspiration for developing a politically engaged epistemology for migration studies. Paraphrasing the revolutionary slogan "action substitutes tears", epistemological and political solidarity with migrants substitutes feelings of shame. In the framework of this chapter, the emancipatory research paradigm of disability research acted as a template for developing four principles for a politically engaged migration studies epistemology. These principles align with efforts "to destabilize the binaries of researcher and researched, focusing instead on the identification or creation of spaces of engagement and proximity, sites of shared struggle and precarity. And they highlight the diverse practices by which mobile subjects negotiate and contest shifting forms of domination and exploitation" (Casas-Cortes et al. 2015).

This is not an easy task and several challenges remain to be addressed. The idea of an emancipatory migration research paradigm, as described above, is merely the first step towards such an ambitious goal, and this publication is not even close to a complete framework for a political epistemology of migration. Such a framework requires the translation of the rich experience of several migration scholars from the field into a systematised and politically informed epistemological agenda that includes the views and needs of migrants. This is a call for researchers who have faced similar dilemmas, emotions, and dead ends in the field to write future publications, and highlights the wish to engage in a research activity compatible with political solidarity with migrants. This embryonic idea and the four principles constitute a set of reflections and a basis for sparking this discussion. It is, after all, our academic and political duty to return to our research subjects the affluence of our research. This is my way to stay loyal to my own political principles and conduct research while performing solidarity with migrants.

NOTES

1. According to Rozakou (2018), the term solidarian is a neologism. In the context of the latest migration flows to Greece, "the solidarian has turned from an adjective to a noun; this grammatical shift signifies the radicalisation of solidarity that took place in austerity-ridden Greece and the flourishment of solidarity. The diffusion of the notion is interrelated with the reconfiguration of the conceptions of the notions of 'social' that has taken place in the country. The expansion of solidarity, and solidarity with migrants in particular, is an essential element of the political content of sociality in this particular historical conjuncture" (ibid., 189). See also Sandberg, Mollerup and Rossi, Chapter 3 in this volume.
2. The hotspot solution was presented by the EU in the spring of 2015 as part of a larger policy push termed the European Agenda on Migration (European Commission 2015). The camps were initially implemented to enable identification, registration, and fingerprinting of arriving migrants, (Antonakaki et al. 2016). However, several researchers have addressed the multifunctional political role of the hotspot approach for the EU migration policy and the management of migratory flows.
3. The EU-Turkey Statement was signed in March 2016 to resolve the 2015 "migrant crisis" and despite critics, it is considered an efficient policy move

that stands out as an exemplar solution for similar future developments (Delcker 2017).

4. It is not my intention here to disregard academic traditions, such as engaged anthroplogy (see for example the introduction in Engaged Anthropology—Views from Scandinavia) with important contributions on similar topics. This chapter aspires positioning itself in the larger community of engaged scholarship.

5. For a thorough discussion on the risks and benefits of publishing migration research results, see Düvell et al. (2010).

6. As in "lateral and anti-hierarchical relatedness [...] in contrast to both hospitality (the dominant cultural code of dealing with alterity) and bureaucratic frameworks of assistance to immigrants and refugees" (Rozakou 2018, 189). My understanding is that political solidarity does not constitute a humanitarian issue that will be resolved when the national state takes into consideration the human rights of migrants according to the liberal principles of the French revolution. Solidarity, in this context, refers to the horizontal gathering of disparate elements that enables the formation of a collective political movement (May 2013).

7. Several insurrectionary events have occurred in the Moria camp since its inauguration, which led to it being finally and totally burnt down in September 2020.

BIBLIOGRAPHY

Ahmed, Sara. 2014. *Cultural Politics of Emotion*. Edinburgh University Press.

Al Jazeera. 2018. "Rare Look at Life Inside Lesbos' Moria Refugee Camp." Retrieved from https://www.aljazeera.com/gallery/2018/1/19/rare-look-at-life-inside-lesbos-moria-refugee-camp.

Antonakaki, Melina, Bernd Kasparek, and Georgios Maniatis. 2016. "Counting Heads and Channeling Bodies: The Hotspot Center Vial in Chios, Greece, Report after Fieldwork." Research Project Transit Migration 2.

Arendt, Hannah. 1951. *The Origins of Totalitarianism*. New York: Schocken Books.

Arendt, Hannah. 1972. *Crises of the Republic: Lying in Politics, Civil Disobedience on Violence, Thoughts on Politics, and Revolution* (Vol. 219). San Fransico, CA: Houghton Mifflin Harcourt.

Athens-Macedonian News Agency. 2019. "Greek Authorities Awaiting Coroner's Report on Asylum-Seeker's Death in Moria, Migration Min. Says." Retrieved from https://www.amna.gr/en/article/323142/Greek-authorities-awaiting-coroners-report-on-asylum-seekers-death-in-Moria--Migration-Min-says.

Badiou, Alan. 2004. "Fifteen Theses on Contemporary Art." *Lacanian Ink* 23.

Balouziyeh, John. 2017. "Dispatch from Moria Refugee Camp: A Crisis within a Crisis." Available at SSRN 3152178.

Barnes, Colin. 2003. "What a Difference a Decade Makes: Reflections on Doing 'Emancipatory' Disability Research." *Disability and Society* 18 (1): 3–17.

Black, Richard. 2001. "Fifty Years of Refugee Studies: From Theory to Policy." *International Migration Review* 35 (1): 57–78.

Burawoy, Michael. 2004. "The World Needs Public Sociology." *Sosiologisk tidsskrift* 12 (3): 255–272.

Casas-Cortes, Maribel, Sebastian Cobarrubias, Nicolas De Genova, Glenda Garelli, Giorgio Grappi, Charles Heller, Sabine Hess, Bernd Kasparek, Sandro Mezzadra, Brett Neilson, and Irene Peano. 2015. "New Keywords: Migration and Borders." *Cultural Studies* 29 (1): 55–87.

Clifford, James, and George E. Marcus. 1986. *Writing Culture: The Poetics and Politics of Ethnography*. Berkeley: University of California Press.

Colson, Nicole. 2017. "Despair in Europe's Refugee Camps." *Green Left Weekly* (1132), 13.

De Genova, Nicolas, ed. 2017. *The Borders of "Europe": Autonomy of Migration, Tactics of Bordering*. Durham, NC: Duke University Press.

Delcker, J. (2017, July 28). "Architect of EU-Turkey refugee pact pushes for West Africa deal." Retrieved March 20, 2021, from POLITICO website: https://www.politico.eu/article/migration-italy-libya-architect-of-euturkey-refugee-pact-pushes-for-west-africa-deal/.

Deutsche Welle. 2019. "Lesbos: Hellish Conditions for Refugees in Moria." Retrieved from https://www.dw.com/en/lesbos-hellish-conditions-for-refugees-in-moria/a-50384674.

D'Haenens, Leen, Willem Joris, and François Heinderyckx. 2019. *Images of Immigrants and Refugees in Western Europe*. Leuven: Leuven University Press.

Düvell, Franck, Anna Triandafyllidou, and Bastian Vollmer. 2010. "Ethical Issues in Irregular Migration Research in Europe." *Population, Space and Place* 16 (3): 227–239.

European Commission. 2015. *The hotspot approach to managing exceptional migratory flows*.

Farkas, Johan, Janick Schou, and Christina Neumayer. 2018. "Platformed Antagonism: Racist Discourses on Fake Muslim Facebook Pages." *Critical Discourse Studies* 15 (5): 463–480.

Fekete, Liz. 2018. "Migrants, Borders and the Criminalisation of Solidarity in the EU." *Race and Class* 59 (4): 65–83.

Foley, Douglas E. 2002. "Critical Ethnography: The Reflexive Turn." *International Journal of Qualitative Studies in Education* 15 (4): 469–490.

French, Sally, and Swain, John. 1997. "Changing Disability Research: Participating and Emancipatory Research with Disabled People." *Physiotherapy* 83 (1): 26–32.

Fujimura, John. 1991. "On Methods, Ontologies and Representation in the Sociology of Science: Where Do We Stand?" In *Social Organization and Social Process: Essays in Honor of Anselm Strauss*, edited by D. Maines, 207–249. New York: De Gruyter.

Galis, Vasilis. 2006. *From Shrieks to Technical Reports: Technology, Disability and Political Processes in Building Athens Metro*. Linköping University Studies in Arts and Science No. 374. PhD thesis, Linköping University, Sweden.

Galis, Vasilis, and Anders Hansson. 2012. "Partisan Scholarship in Technoscientific Controversies: Reflections on Research Experience." *Science as Culture* 21 (3): 335–364.

Gordon, Eleanor, and Henrik K. Larsen. 2020. "'Sea of Blood': The Intended and Unintended Effects of the Criminalisation of Humanitarian Volunteers Rescuing Migrants in Distress at Sea." *Disasters*.

Gray, Breda. 2008. "Putting Emotion and Reflexivity to Work in Researching Migration." *Sociology* 42 (5): 935–952.

The Guardian. 2019. "Oxfam Condemns EU over 'Inhumane' Lesbos Refugee Camp." Retrieved from https://www.theguardian.com/world/2019/jan/09/oxfam-criticises-eu-inhumane-lesbos-refugee-camp-moria.

The Guardian. 2020. "Moria Is a 'Hell': New Arrivals Describe Life in a Greek Refugee Camp." Retrieved from https://www.theguardian.com/global-development/2020/jan/17/moria-is-a-hell-new-arrivals-describe-life-in-a-greek-refugee-camp.

Haraway, Donna. 2001. "Situated Knowledges: The Science Question in Feminism and the Privilege of Partial Perspective." In *The Gender and Science Reader*, edited by M. Lederman and I. Bartsch, 169–188. London: Routledge.

Honig, Bonnie. 2009. *Emergency Politics: Paradox, Law, Democracy*. Princeton: Princeton University Press.

Jørgensen, Martin Bak. 2012. "Categories of Difference in Science and Policy—Reflections on Academic Practices, Conceptualizations and Knowledge Production." *Qualitative Studies* 3 (2): 78–96.

Lind, Jacob. 2020. *The Politics of Undocumented Migrant Childhoods: Agency, Rights, Vulnerability*. Doctoral dissertation, Malmö University.

Lundberg, Anna, and Emma Söderman. 2015. "Reflections on the Right to Health." In *Social Transformations in Scandinavian Cities: Nordic Perspectives on Urban Marginalization and Social Sustainability*, edited by E. Righard, M. Johansson, and T. Salonen, 251–264. Nordic Academic Press.

Lundberg, Anna, and Michael Strange. 2017. "Who Provides the Conditions for Human Life? Sanctuary Movements in Sweden as Both Contesting and Working with State Agencies." *Politics* 37 (3): 347–362.

Lury, Celia, and Nina Wakeford, eds. 2012. *Inventive Methods: The Happening of the Social*. Routledge.

Mackenzie, Adrian. 2012. "Set." In *Inventive Methods: The Happening of the Social*, edited by C. Lury and N. Wakeford, 219–231. London: Routledge.

Marres, Noortje. 2012. "The Experiment in Living." In *Inventive Methods: The Happening of the Social*, edited by C. Lury and N. Wakeford, 76–95. Routledge.

May, Todd. 1994. *The Political Philosophy of Poststructuralist Anarchism*. University Park, PA: The Pennsylvania State University Press.

May, Todd. 2013. "Humanism and Solidarity." *Parrhesia* 18: 11–21.

Mezzadra, Sandro. 2011. "The Gaze of Autonomy: Capitalism, Migration and Social Struggles." In *The Contested Politics of Mobility: Borderzones and Irregularity*, edited by Vicki Squire, 121–142. Routledge.

Millner, Naomi. 2011. "From 'Refugee' to 'Migrant' in Calais Solidarity Activism: Re-staging Undocumented Migration for a Future Politics of Asylum." *Political Geography* 30 (6): 320–328.

New York Times. 2018. "'Better to Drown': A Greek Refugee Camp's Epidemic of Misery." Retrieved from https://www.nytimes.com/2018/10/02/world/europe/greece-lesbos-moria-refugees.html.

No Border Kitchen. 2019. "In Memory and Rage." Retrieved from https://noborderkitchenlesvos.noblogs.org/post/2019/01/.

Nordstrom, Carolyn, and Antonius C. Robben, eds. 1995. *Fieldwork under Fire: Contemporary Studies of Violence and Culture*. University of California Press.

Oliver, Mike. 1992. "Changing the Social Relations of Research Production?" *Disability, Handicap and Society* 7 (2): 101–114.

Pittaway, Eileen, Linda Bartolomei, and Richard Hugman. 2010. "'Stop Stealing Our Stories': The Ethics of Research with Vulnerable Groups." *Journal of Human Rights Practice* 2 (2): 229–251.

Pitzer, Andrea. 2017. *One Long Night: A Global History of Concentration Camps*. Hachette, UK.

Reuters. 2019. "'Moria Is Hell': Asylum Seekers Protest Conditions at Greek Camp." Retrieved from https://www.reuters.com/article/uk-europe-migrants-greece-lesbos-protest-idUKKBN1WG3WA.

Robben, Antonius. 1995. "The Politics of Truth and Emotion among Victims and Perpetrators of Violence." In *Fieldwork Under Fire: Contemporary Studies of Violence and Culture*, edited by C. Nordstrom, and A. C. Robben, 81–103. University of California Press.

Rozakou, Katerina. 2017. "Solidarity# Humanitarianism: The Blurred Boundaries of Humanitarianism in Greece." *Etnofoor* 29 (2): 99–104.

Rozakou, Katerina. 2018. "Solidarians in the Land of Xenios Zeus: Migrant Deportability and the Radicalisation of Solidarity." In *Critical Times in Greece: Anthropological Engagements with the Crisis*, edited by D. Dalakoglou and G. Agelopoulos, 188–201. Routledge.

Scott, Pam, Evelleeen Richards, and Brian Martin. 1990. "Captives of Controversy: The Myth of the Neutral Social Researcher in Contemporary Scientific Controversies." *Science, Technology and Human Values* 15 (4): 474–494.

Sillitoe, Paul, ed. 2016. *Indigenous Studies and Engaged Anthropology: The Collaborative Moment*. Routledge.

Spivak, Gayatri Chakravorty. 1988. "Can the Subaltern Speak?" In *Marxism and the Interpretation of Culture*, edited by C. Nelson and L. Grossberg, 271–313. Urbana, IL: University of Illinois Press.

Ticktin, Miriam. 2014. "Transnational Humanitarianism." *Annual Review of Anthropology* 43: 273–289.

Ticktin, Miriam. 2017. "A World without Innocence." *American Ethnologist* 44 (4, November): 577–590

Titley, G. 2019. *Racism and Media*. Sage.

Turton, David. 1996. "Migrants and Refugees: A Mursi Case Study." In *In Search of Cool Ground: War, Flight and Homecoming in Northeast Africa*, edited by Tim Allen, 96–110. London: Africa World Press.

Tyler, Imogen. 2006. "'Welcome to Britain': The Cultural Politics of Asylum." *European Journal of Cultural Studies* 9 (2): 185–202.

Vandevoordt, Robin. 2019. "Subversive Humanitarianism: Rethinking Refugee Solidarity Through Grass-Roots Initiatives." *Refugee Survey Quarterly* 38 (3): 245–265.

Van Liempt, Ilse, and Veronika Bilger, eds. 2009. *The Ethics of Migration Research Methodology: Dealing with Vulnerable Immigrants*. Sussex Academic Press.

Van Liempt, Ilse, and Veronika Bilger. 2018. "Methodological and Ethical Dilemmas in Research Among Smuggled Migrants." In *Qualitative Research in European Migration Studies*, 269–285. Cham: Springer.

Venkatesh, Sudhir Alladi. 2013. "The Reflexive Turn: The Rise of First-Person Ethnography." *The Sociological Quarterly* 54 (1): 3–8.

Žižek, Slavoj. 2008. *Violence: Six Sideways Reflections*. London: Profile books.

CHAPTER 8

Emotional Introspection: The Politics and Challenges of Contemporary Migration Research

Ninna Nyberg Sørensen

INTRODUCTION

Hands that usually write meticulous field notes lie still on the table in the room where a seasoned researcher—recognised among peers for his fieldwork in places marked by conflict—has agreed to meet and share his experiences. Seconds ago, the same hands were eagerly gesturing to illustrate a compelling experience in the field. He now tells me about an encounter with a group of recently arrived, internally displaced women with terrible pasts. While narrating details about their escape from a violent attack by an armed group terrorising their hometown, one of the women suddenly placed her child in his arms and begged him to take it to Europe, "so that the child at least may have a chance". Her words—"there's nothing left to be done here, we're finished"—torment his memory. Making a shift from a passionate researcher to a human being

N. N. Sørensen (✉)
Danish Institute for International Studies (DIIS), Copenhagen Ø, Denmark
e-mail: nns@diis.dk

© The Author(s) 2022   195
M. Sandberg et al. (eds.), *Research Methodologies and Ethical Challenges in Digital Migration Studies*, Approaches to Social Inequality and Difference, https://doi.org/10.1007/978-3-030-81226-3_8

momentarily stripped of his defences, the still hands and empty gaze reveal how emotionally disturbing fieldwork can be.

"It's the combination", he says, referring to the exposure to human suffering, personal expectations of making a difference, and very narrow opportunities to do so. "It's just so incredibly insurmountable". Unlike when he was younger and convinced that his work could make a difference to the quality of life of his interlocutors, the accumulation of instances of "not being able to do anything" occasionally fills him with hopelessness, desperation, and, in his eyes even worse, cynicism.

Scholars who conduct field research among people in difficult circumstances are often confronted with immense hardship and ethical complexities. This is notably the case in contexts where gross inequality, poverty, violence, and physical vulnerability abound, and protective social infrastructures are inadequate and unevenly accessible (Plüg and Collins 2018). As qualitative research is predicated on establishing trusting relationships with groups and individuals in local environments, field-based research in fragile situations and insecure places, almost by definition, means putting ourselves not only physically at risk, but also emotionally on the line (Goldstein 2014). Risk of emotional distress certainly applies to migration researchers who increasingly encounter traumatic events and narratives associated with involuntary immobility or clandestine travel across borders in attempts to escape poverty, violence, or persecution. The migrant experiences that often leave the heaviest imprint are those of migration gone terribly wrong: stories of not succeeding in crossing the border; of being intercepted and forcefully returned along the routes; of, after years of living undocumented lives abroad, being repatriated from refugee camps or becoming "victims of human trafficking"; or stories told by relatives searching for loved ones who disappeared before reaching their destination. Failure to connect such experiences to the effect they have on our work and personal lives will inevitably affect the way we view the world and ourselves in it and, by implication, our ability to carry on producing sound research.

Anthropologists and others applying ethnographic methods have an ample vocabulary for how we affect or impose ourselves on the field. When it comes to how experiences in the field affect us, the conversation gets quieter (Kristensen 2017). This fragmentation is partly a result of the unresolved role of emotions in research (Fleetwood 2009), partly a consequence of a lack of institutional structures that seriously deal with these matters, and partly an effect of broader patterns of unequal power relations between researchers, the people whose lives we wish to engage with, learn from and document, and the local brokering researchers engaged, without whom much research could not be carried out (Abedi Dunia et al. 2019; Abedi et al. 2020). And it is by no means an easy conversation to engage in; on the few occasions when we expose the often-brutal reality our research entails, it can occasion serious digestive problems among ethics committees, local partners, colleagues, and, not least, ourselves due to personal ambivalence about the relevance and "place" of revealing our emotions publicly, with or without good cause (McLean 2011). It is therefore no surprise that migration researchers occasionally face emotional challenges comparable to those encountered by colleagues in fields such as peace and conflict studies, or by staff working for humanitarian organisations, among whom high rates of anxiety, depression, compassion fatigue, secondary traumatic stress, or vicarious trauma are found (Grimm et al. 2020). We just rarely talk about it and tend to ignore symptoms of compassion fatigue or vicarious trauma when our spirits get low. Some, then, go down with "stress".

Over the past ten years, the question of physically "surviving fieldwork" has become a central concern in research (Shiram et al. 2009; Felbab-Brown 2014; Theidon 2014; Grimm et al. 2020; Mac Ginty et al. 2021). In response, many research institutions have taken steps to reduce fieldwork risks and enhance safety. Safer research is generally understood to involve predeparture preparation and, when deemed necessary, some sort of hostile environment awareness training (so-called HEAT courses). The safety of interlocutors is generally monitored by committees overseeing compliance with core ethical principles such as informed consent, "do no harm", and data protection and storage (Mackenzie et al. 2007; Hugman et al. 2011). Debriefing structures are sometimes set up, offering opportunities to offload shattering experiences, inspired by those implemented by humanitarian actors (Martin-Ortega and Helman 2009). While no one can be against preparing researchers for safety hazards, critical voices argue that one-sided attention to physical security risks encouraging an "us versus them" attitude (Lake and Parkinson 2017) that, furthermore, may contribute to increasing the distance between researchers and our interlocutors (Grimm et al. 2020). Another problem with the one-sided focus on physical safety pertains to the possible contribution to undifferentiated, and at times exaggerated, negative stereotypes of dangerous settings (Woon 2013), contributing to the creation of a "no-go world" of global distancing and endangerment (Andersson 2019). These concerns resonate well with Sara Ahmed's suggestion that judging places and bodies as suspicious, dangerous, or something to be feared may have fatal consequences for those inhabiting them (Ahmed 2014). Awareness of these critical issues does not exactly make the conversation easier.

Attempts to promote safer field research attuned to the reality of the social sciences have been slow to incorporate emotional challenges and well-being, including not only that of contracting researchers institutionally located in the Global North but also of facilitating researchers and other participants in the Global South (Môcnik 2020). Acknowledging that emotional well-being is closely connected to physical security, and that neither of these concerns can be limited to the contracting researchers only, this chapter aims at discussing the emotional challenges involved in contemporary field research among migration researchers. The chapter highlights the contexts and institutional structures surrounding our research, the new opportunities for conducting digital research along with the associated effects of digital availability, and what can be done *prior to, during*, and *after* going to the field to maintain emotional engagement while avoiding occupational hazards of compassion fatigue, burnout, or secondary traumatic stress among all those involved.

Before embarking on these issues, I briefly present the methodology on which my reflections are based and describe the current context for conducting migration research. Drawing on fieldwork experiences shared by colleagues, I then present some of the emotional challenges encountered. I conclude by highlighting some of the practices that have proven useful for recognising emotions as intrinsic to field encounters while simultaneously helping to reduce the negative effects and personal costs associated with fieldwork. Since symptoms of vicarious trauma and burnout among researchers are well described in the psychological literature, my emphasis is on the participating researchers' personal experiences, explanations, and suggestions for ways ahead, rather than on the occupational hazards resulting in headaches, emotional extraction, sleep problems, drained immune systems, altered perceptions of reality, self-blame, cynicism, and tunnel vision that most have experienced in one form or the other.[1]

A Note on Methodology

After years of institutional preoccupation with research ethics and the physical safety of researchers and their interlocutors, I have discovered that a "loss of heart" among colleagues often had as much to do with being exposed to seemingly hopeless situations as with being provoked by working in dangerous settings. Before I got there, I took a detour around occupational stress, which seemed to be prevalent among colleagues, especially PhD students, who are almost expected to succumb to stress during their doctoral research. There is much to ponder here, which, beyond questions of the "illegitimacy" of feelings among researchers and inadequate training in dealing with them, also includes the generalised belief—whether true or not—that emotionally "tough" researchers stand a better chance in the struggle for (limited) future permanent positions. Thus, it may be regarded as more legitimate to give in to the pressures of academia than to the emotions fieldwork may have provoked.

The empirical basis for the experiences related here stems from two years of dialogue with colleagues and more structured interviews with ten selected scholars. The researchers contributing their experiences are diverse in terms of gender, academic trajectory, national background, and institutional location. Only one researcher, however, is physically located in the Global South, whereas the rest are attached to Danish or other European research institutions. Their disciplinary backgrounds vary, spanning anthropology, sociology, geography, history, psychology, and development studies, with different traditions of including/excluding emotional aspects in the research process. All have conducted one or more long-term periods of fieldwork in insecure places among people subjected to extreme hardship related to some form of migration, be it among internally displaced persons (IDPs); people fleeing poverty or war involving genocide; people trapped in refugee camps, asylum and detention centres; or smuggled and trafficked migrants. While many of the migrants and refugees interviewed by my conversation partners have managed to find better and more secure lives over the years, others have ended up in prison, died along the routes, or continue to eke out an existence. The disastrous consequences of deterrence-based migration policy and practice—that deliberately ignores the plight of migrants in distress in order to dissuade others—are not limited to any particular region. The participants' field experiences stem from Africa, the Middle East, Asia, and Latin America, as well as from sites within or bordering Europe and the

United States. This diversity is not totally random. Several are present or former colleagues; others are people I have met at conferences or through other forms of international research collaboration. While I considered approaching researchers outside my personal networks to avoid "insularity", upon reflection this probably would have led to less focused and less intimate conversations. I already knew about (some) of the emotional challenges the participants had faced; they knew me well enough to feel confident talking about less "heroic" moments of their careers.

All participants were given advance information about the aim of my project and sent a rough thematic "guide" about issues of interest, including specific, demanding fieldwork incidents, whether we are prepared for such incidents to occur, how we tackle emotional challenges during and after fieldwork, and the degree of institutional support in pre-fieldwork training, supervision and—when necessary—psychological counselling during and after completed fieldwork. Everyone was encouraged to make constructive suggestions about what we can do better, both as individuals and as research institutions. Finally, more established researchers were asked to reflect upon whether their field has become more emotionally challenging over the years, an issue I find particularly important in efforts to avoid over-simplification (e.g. that younger researchers, due to a lack of experience, are more prone to emotional hardship or that burnout among older researchers is necessarily due to overexposure).

Our exchanges lasted from one to two hours. Some took place in my or their offices, others in places outside our working environments. Three conversations were held virtually. On all occasions, we discussed the purpose of dwelling on emotional hardship in research before oral informed consent to participate was granted. All conversations were taped and fully transcribed by me. The participants were offered the opportunity to develop statements they found needed further explication than in the transcripts. They have also been offered the opportunity to see a draft of this chapter before publication. Finally, all have been granted anonymity and no names therefore appear in what follows.

Despite my focus on personal experiences, several participants reflected not only on their own but also on situations affecting local brokering partners, be they interpreters or fellow researchers contracted for data collection. Instead of being "oblivious to privilege and positionality" (Abedi Dunia et al. 2019), most were painfully aware of having had access to insurance or counselling structures not available to local partners. Indeed, frustrations over unequal working conditions have, in some cases, exacerbated the emotional strain during fieldwork. Such frustrations have also been found among researchers from the Global South who, for instance, study for their PhD degrees at universities in the North and from that position conduct field research in their countries of origin (Kalinga 2019; Turner 2010).

CONDUCTING MIGRATION RESEARCH AT THE PRESENT MOMENT

Migration research is a diversified field concerned with the drivers, processes, and outcomes of proactive ("voluntary") and reactive ("forced") manifestations of human mobility. Those involved in the field increasingly agree that the distinction between economic and political forces motivating migratory movements is difficult to uphold (Richmond 1994; Stepputat and Sørensen 2014). Similar critical reflections on migration as exceptional human behaviour have begun to take shape (Shah 2020), just as studies of immobility have gained traction in migration research (Carling 2002; Lubkemann 2008). Human mobility, especially over long distances, has always been subject to contestation. Studying these processes ethnographically in real time has more often than not involved dilemmas of privilege and disadvantage, attention to asymmetric and unequal power relations and, in the best of cases, concerted reflection on ethical challenges along with methodological and political investment in research (Castellanos 2019).

Many challenges befall scholars who explore contemporary migration phenomena. On top of the power dynamics between researchers and those whose lives are at stake, the representation, dissemination, and discussion of research results increasingly take place in politicised fora. Public discourse and the media play a major role in framing perceptions of migration. When framed as crisis, migration experiences tend to acquire a specific temporal dimension, largely disconnected from history and related events, that not only directly shapes policies but also favours some research results over others (Menjívar et al. 2019). Indeed, researchers have far better access to policy circles and media outlets than disadvantaged interlocutors, but what we contribute and how that contribution resonates with politics vary according to the prevailing narrative of migration. Or, put differently, conducting migration research may be perceived as more challenging during times when research results have little resonance among political decision-makers and the general public, when the lines between "for" and "against" migration are more sharply drawn, or when humanitarian considerations become subordinated to wishes for control and restraint. Thus, emotional experience cannot be separated from the historical moment and social environment in which is embedded (Åhåll 2018).

Crisis rhetoric and images of "unprecedented flows" have led to new restrictions on human mobility, manifested in ever stricter border control mechanisms or in what Ruben Andersson (2019) has called a "post-humanitarian moment of harsher deterrence". Deterrence mechanisms include the extension of border surveillance and cooperation between coast and border guards with migrant-sending countries (de facto illegalising not only the right to *enter* a given country but also to leave your own), the use of suffering as a means of deterrence, the deployment of security forces in border and desert areas (to perform "pushbacks" and "pullbacks"), the outsourcing of migrant detention and removal to private security companies, and the criminalisation of assistance (Sørensen and Plambech 2019; Distretti 2020).

Alongside the geopolitical changes affecting migrants' and asylum seekers' travel, transit, and arrival conditions, there are questions of extortion, violence, abuse, enslavement, and, in the worst cases, disappearance or death along the routes that demand attention (Sørensen and Huttunen 2020). Such changes and developments are reflected in the places where researchers conduct fieldwork, including insecure border areas or other hostile environments. They are also reflected in the situations in which fieldworkers find their interlocutors. As global inequality is currently manifested in unequal access to mobility and fundamental human rights, many of these situations involve indefinite positions of "not arriving" and "not becoming" (Khosravi 2018), meaning that many migrants and refugees are found in seemingly hopeless situations. It is therefore no exaggeration to say that migration research is currently challenged by increasing media attention to, on the one hand, imagined or real increases in the numbers of migrants arriving at the borders of reluctant hosts, and, on the other hand, traumatic migrant experiences of dramatic journeys, pushbacks, detections, detentions, deportations, and gross human rights violations along the way.

With so much insecurity and human suffering at stake, the central ethical and methodological challenges in undertaking migration research demand renewed attention. Research involving refugees and IDPs has acknowledged the politically complex, difficult, and often insecure places in which fieldwork among vulnerable participants is carried out. This has occasioned ethically reflexive calls for ways to move beyond "do no harm" as a standard for research design and conduct (Mackenzie et al. 2007). However, there has been less concern with the emotional implications and effects of conducting research in insecure places among migrants in extremely difficult situations.

Nonetheless, a small but growing emphasis on the ethical challenges and emotional politics involved in conducting fieldwork on migration-related topics in times of intensified border controls is beginning to emerge, often building on personal experiences in the field and published in blog posts. Writing on secondary trauma, Mary Bosworth notes that her research in immigration detention centres is "unsettling" and sometimes "overwhelming", leading to psychological effects such as sleeplessness, anxiety, and palpitations. While Bosworth depicts these emotions as hers to manage and accepts them as an inescapable part of where her work as an experienced criminologist takes her, she finds it crucial to share accounts of challenges and how to overcome them with doctoral students and postdocs, who might be entering the field for the first time (Bosworth 2017). From the position of a younger researcher embarking on field research, Tamara Last describes how conducting fieldwork on border death led to symptoms of depression and the development of PTSS in herself and among her co-workers. Although such experiences have a high personal cost, when subjected to critical (auto-)reflection they may lead to methodological suggestions for reducing vicarious trauma stemming from emotionally disturbing research, including ideas for integrating effective monitoring and intervention protocols into the research design from a very early stage (Last 2020; see also Môcnik 2020).

Attention to the role of emotions in fieldwork provides a pathway to appreciate the imprint of personal relations and feelings that shape research outcomes and the knowledge produced. We go to the field to get a more nuanced understanding of local and geopolitical processes, to enhance the visibility of grounded and alternative visions, and hopefully thereby become capable of decentring assumptions that are taken for granted. "Emotional fieldwork", Chih Yuan Woon argues, "opens up the researcher to different emotional engagements and connections with his/her respondents, which in turn allows for critical reassessment of issues pertaining to danger, ethics and responsibility" (Woon 2013, 32).

However, even if such insights provide a powerful argument against positivist research ideals and effectively deconstruct objectivity and distance in favour of emotional engagement, they also hint at the somewhat false dichotomy between the physical and emotional safety of field-working researchers. This has been a recurring theme in the conversations that inform the following discussion.

FIELD-RELATED EXPOSURE, INSTITUTIONAL DRIVERS, AND EMOTIONAL VULNERABILITY

Field-based research often takes us to places that are unimaginable before arriving and unpredictable while there. While predominantly presented as a professional practice, engaging in fieldwork often turns out to be a physical, personal, and emotional experience (Fleetwood 2009). This also applies to fieldwork conducted in the countries we reside in—for example when conducting research in refugee reception or migrant detention centres—with the difference that the researcher can return to her or his home on a daily or weekly basis and maintain direct social contact with family, friends, and colleagues. However, the ability to reach out is not necessarily directly connected to physical distance between "home" and "field".

Insecure Places and Hopeless Situations

In terms of field-related, emotionally distressing experiences, the participants in this study distinguished between *anxiety provoked by insecure places* and *discouragement provoked by exposure to hopeless situations*. Talking about both proved equally transgressive for most. Anxiety leading to emotional distress was partly experienced as related to physical dangers abroad such as the risk of falling victim to violence, kidnapping, or traffic accidents, whereas anxiety grounded in the difficulty of being able to distinguish what is dangerous in unfamiliar surroundings constituted another group of factors. The daily unpredictability coupled with doubts about whom to trust, feeling monitored by authorities or parallel criminal structures, or feeling manipulated by gatekeepers, were all seen as contributing to distress in the field and the development of stress symptoms upon return, leading to sick leave (three cases) and, in the worst case, vicarious trauma upon return (one case) or even years after leaving the field (another case).

Discouragement provoked by exposure to desperate and hopeless situations was easier to acknowledge on a personal level but equally difficult to talk about. "You cannot afford to succumb to the suffering of others. They have to stay and live with it all the time", said one participant. While doing research among deported migrants in the Global South, an experienced researcher noted: "Well, they live there all the time, and I can leave. So, I have many discussions with myself that are both personal and political. Can I allow myself as a white, middle-class person to say that I do not dare be in a certain place that others are prevented from leaving?".

Many did not receive much guidance before embarking on their first fieldwork. They were familiar with qualitative interview techniques, participant observation, and so on from their basic academic training; but had never spoken to their PhD supervisors or senior colleagues about the potential emotional strain of conducting fieldwork. Along with the lack of preparation for the emotional impact of fieldwork on their well-being, younger colleagues also related that they did not have much contact with their supervisors during often-lengthy fieldwork periods. Many, moreover, perceived that it was expected of them to have as little contact as possible with their personal social networks while abroad, to facilitate a deeper engagement with the places and people among whom they worked. Finally, one participant mentioned that supervisors involved in collaborative projects may become vicariously traumatised by the research themselves, which creates dysfunctional supervision and support structures.

Emotional Rewards

Although most could talk about emotionally challenging situations in the field, hardly any started the conversation there. On the contrary, participants commented that they had felt enriched, and naturally also saddened, by what their interlocutors had gone through, but even so, they felt professionally and personally rewarded by the, often intense, meetings with other human beings. "People always expressed contentment with my being there, of being asked to tell their story", a researcher explained. In his experience, many hard stories led to some form of relief, for the person telling the story as well as for the researcher tasked with making the story known. The latter, of course, places an immense burden on the act of bearing witness (Appelbaum 2008) that may manifest in the process of analysis and writing. To this researcher, however, allowing himself to be emotional in the field and to not be afraid of his personal vulnerability was, furthermore, an essential methodological necessity: "I cried during several interviews, and that was what opened up ... in the moment people saw my emotional reaction, they threw away their reservations and opened up completely". Allowing ourselves to feel hurt during fieldwork is therefore not only essential to maintaining a healthy emotional self but also a methodological tool for gaining "access".

Another researcher, who for much of his lengthy career has worked with violent conflict, IDPs, and international refugees, expressed a slightly different dynamic: "After all, we are talking to the survivors, who are often capable of turning the past into some sort of heroic narrative of how they managed the situation". In his view, then, the reward of fieldwork consists in providing interlocutors with a space to create meaning and take on an active role in an otherwise horrific situation.

Such spaces are not one-way, but often reflect on the researcher's emotional well-being: "I think that I grew personally from working with vulnerable people", a mid-career researcher underscored. "And this is not because I compare my own situation to theirs. It is rather because it is in the meeting between human beings that you gain energy ... you know, the opposite of compassion fatigue". She further explained:

> I believe it is all about accepting that you move through different spaces but maintain the same ethics in them. It is a form of relativism where you accept that we have very different conditions, but basically we are the same. And always respect the persons you work among, never say no, but always accept what someone offers you, you know, somehow being able to show that you are able to see the resources people have. Even in the darkest of situations.

Others nevertheless ended up pointing to the ambivalent place of emotions in research and the institutional structures that prevent taking emotions seriously into account. "Are you really allowed to have feelings in research?" was a frequently asked question, as were statements such as "I believe emotions continue to be illegitimate in research environments" or "ideas of objectivity still prevail and prevent us from bringing our personal emotions to the fore". Giving the discussion a gendered twist, one could further argue that emotions remain associated with the personal, the body, and the feminine, whereas the objective, the mind, and the masculine continue to prevail, even in the sciences that are supposed to be "social" (Åhåll 2018). Younger colleagues pointed to a dire lack of institutional attention to emotional stress both prior to, during, and to a lesser extent after completing their fieldwork, while postdocs searching for permanent employment expressed a fear that showing "emotional weakness" could be used as a sorting mechanism when filling new positions. "I feel that you gain recognition by daring to go to difficult places", one discussion partner stated, whereas another felt that "to be emotional is to be vulnerable and vulnerability is not viewed as a resource when it comes to hiring permanent staff". Younger researchers therefore felt that they could talk to colleagues on their own level, but not necessarily to their supervisors or other senior researchers in the institutions to which they are (temporarily) attached.

Conducive Institutional Cultures?

Institutional structures reflect power relations with respect to project approvals, grants, and, ultimately, promotions along the career track. Apart from how they are structured, institutional cultures are co-created by how researchers represent themselves and interact with each other. Mirroring the "Bang-Bang effect" among war reporters and photographers (Marinovich and Silva 2000); Kimberly Theidon (2014) refers to the practice of sharing who has seen the goriest scenes (the most battered bodies, the heaviest rain of bullets, or other dangers) as the "horror index" within peace and conflict studies. The stories I have encountered among migration researchers have not necessarily been gory but nevertheless may have contributed to generating what I, for lack of a better concept, term the "Indiana Jones syndrome".

The following incident took place at a party among peers in 2018. The planning committee had done its utmost to create a selection of themed bars. One such themed bar was inside a tent, furnished with carpets, cushions, and dimmed lighting. Malinowski's tent on the Trobriand Islands? A refugee camp facility? Once inside, all attendees were asked to write down a fieldwork experience, either experienced by themselves or by someone known to most. The notes were collected in a bowl. The following game consisted of guessing who the experience referred to: "Who was once hanging on a cliff after a landslide", "who unknowingly attended a party with narco lords", or "who was robbed at gunpoint outside his field residence", and so on. Conspicuously absent were questions such as "who was left sleepless for months after working in a refugee camp" or "who was on sick leave after carrying out fieldwork among human smugglers". People were surprisingly good at guessing whom each incident referred to, a fact I interpret as indicative of the rather high street value such stories have in research environments. But, also, telling of the fact that emotional stress management can take many forms, including performance, good company, and shared laughter. This, then, obviously leads to the question of the extent to which institutional cultures limit or enable attention to research-related emotional distress.

Not everyone buys into the "Indiana Jones syndrome". One mid-career researcher felt "totally misplaced" in settings cultivating this kind of "boyish attraction to danger". After years of working among asylum seekers and immigrants in prisons or detention centres, it was her experience that some male colleagues are attracted to what they believe is work among dangerous inmates:

> I find that way of talking about the people we work with quite disturbing, enormously 'other-making'. It is making people far more dangerous than they are, or ever could be, and it results in a way of articulating otherness as something you cannot grasp unless you have been there, done drugs with your interlocutors, shared their weapons.

When asking if she had ever felt emotional distress during fieldwork, she elaborated: "While some of the people I've worked with have been convicted of violence or done other terrible things, only defining them by their criminal act and not taking other facets of their lives into account, including *their* boyishness, the everyday and the adventurous, the poetic and the inspirational, all the other things that go into a person … to omit that disturbs me".

The point here is not to distinguish "hard-skinned" researchers from emotionally reflexive ones, nor to manshame male colleagues, or accuse feminist killjoys, and certainly not to argue for cancelling future parties and opportunities to relieve anxiety-provoking experiences. A good laugh among peers has often proved an effective way to start deeper and relevant discussions about working conditions. However, continuously casting the researcher as adventurous and daring (to not say self-promoting or self-aggrandising) contributes to upholding unhealthy institutional cultures as well as the co-creation of global distancing and no-go zones.

Contagious Field Sites, Empathy and Emotions

An early career researcher conducting fieldwork in asylum centres connected the field site, the personal and the political in a different way: "In many ways, asylum centres are ideal places for conducting fieldwork. Nothing much is happening, most of the time it is pretty boring, and people have plenty of time to talk with you", she explained. But then she continued:

> However, there is a sensory dimension connected to walking around in this boredom. The feeling that everything has stalled but meanwhile your interlocutors could receive a letter from the immigration authorities any day that changes everything. That so much is at stake ... it somehow becomes contagious ... that everything can go wrong [with the asylum application process] any moment ... I found it immensely stressful.

After ending a year of fieldwork, she fell apart. "When everything is so, yes, just shitty, and people are doing so badly and this becomes the norm, well then it becomes extremely difficult to leave the field. Because everything that I find problematic in relation to the handling of asylum seekers is somehow confirmed by my departure, right? And well, I just became very sad". The combination of exposure to human suffering and the "sorrow of parting" (Parvez 2018) simply became too much to bear.

Another researcher turned the question of emotional distress upside-down by asking: "Wouldn't it be terribly disturbing if you were not affected? If you were not emotionally distressed? To see all that suffering without any reaction. Isn't that in fact a sign of psychopathy?" In her definition, watching and navigating human suffering undisturbed, without letting yourself be touched by it, would constitute a horrific fieldwork ideal. As emphasised by Kasper Hoffmann (2014), the proposition that fieldworkers can or should strive to set their personal histories aside—and thereby divest themselves of their values and prejudices in order to attain an objective understanding of their research subjects—is nothing short of an impossible empiricist dream.

The Process of Writing

Equally important, participants described how feelings of emotional distress "never end" but continue or reappear in the process of data processing, analysis, and writing: "I found it really difficult to get back to the taped interviews and read the transcriptions. In fact, I had to have another person transcribe the recordings, because I could hardly stand listening to them, even a year after my return from the field", a PhD researcher stated.

> It is when I get back to the data that I feel vulnerable. And to this day I have a hard time sitting down to listen to the recordings ... it reminds me of how embedded I was and how focused I was on taking physical care of myself, you know: being able to turn the car 180 degrees in two seconds and flee and whatever you learn during those HEAT courses, while I did not consider the emotional damage that feelings of constant insecurity inflicted on me.

In her case too much, or too one-sided, institutional attention to physical security prevented her from grasping her emotional vulnerability.

When field notes or collected archival material contain traumatic human experiences, writing may become insuperable. The collision of pressure to disseminate, institutional demands around formats, and personal expectations contributes to the unbearable darkness of the writing process. Another PhD researcher commented: "Before I was just part of a research project, but you get the sense, well definitely the intense sense, that you need to publish. But here the emotional struggles became a block". In attempts to write she found it impossible to write "objectively", and objective was the standard she was encouraged to write by. "I had to sort of pull all the feelings out of it. And that was very difficult to do, and in the end also very unsatisfying".

A more experienced researcher described how hopelessness, despite years of engaging with the same field, crept under her skin: "I was trained to look for agency, and I have written about agency, but these days migrants are deported anyway". Thus, emotional distress related to writing is not limited to a possible traumatic experience of revisiting field notes and recordings. It is also embedded in the increasingly politicised environment circumscribing our research: "It is my experience that the migrants I work among have an increasingly hard time", she further explained. "When I started out 15 years ago, my interlocutors were granted residence status, some felt lonely and longed for somebody to talk to. And there I was. Today they can be deported at any time ... and it makes me feel powerless. To be witness to this development. What difference does it make that I write about it? Because their situation just gets worse and worse, across the board". To underscore the pervasive dilemmas related to writing she stated: "With the changing political situation I feel, well, that it doesn't help, in this political environment, to reveal in writing that the people I work with have agency, because it will only be used against them".

Digital Availability

Whereas digital technologies have expanded the ways research is carried out, including web-based interviewing and other forms of "netnography" (Kozinets 2015), it has also affected the relationship between researchers and research subjects. The fact that online sociality is multidirectional and increasingly accessible to the people we work with means that it is no longer the sole privilege of the researcher to reach out to interlocutors, as they are now available to interlocutors long after a research project ends. While this certainly democratises research processes and moves research towards more participatory and inclusive designs, it potentially poses an additional stress factor to those already mentioned above, especially if there is so little you can do to change the often dire circumstances our interlocutors find themselves in. As one senior researcher commented:

> During the past few years, I have experienced a tremendous level of stress from constantly being in contact with my interlocutors. Hearing about their situation all the time, through WhatsApp, Messenger, phone calls. I constantly negotiate with myself whether I have to be on fieldwork all the time. What happens if I decide not to? I cannot afford to withdraw for a year and then go back when it suits me and say 'hey, let's resume contact', while they have had a miserable year.

A doctor's order to disconnect from communication with interlocutors during a three-month leave recharged her batteries enough to resume contact. Others talked about the cumulative effect of not handling their field and maintaining contact as well as previously, partly due to the perception that migrants and refugees face ever harsher conditions, partly because their own ability to make a difference has narrowed.

Conclusion and Ways Ahead

You will be changed by your research; that is one of the legacies. No, not necessarily turned into one of the walking wounded, but changed in ways that may not be readily or immediately apparent. The awareness of this – and of how the changes manifest across time and space – can make a difference while we conduct our research and when we return from our field sites and sit down to write. (Theidon 2014, 3)

This chapter has explored the difficulties of practicing emotions in the field and in research in general. To that end, I have outlined examples from the research participants' professional trajectories, as well as their reflections on these. One of the issues pertaining to fieldwork is that we never seem to have as much control over what is going on around us as we would sometimes like. Rather, we pretty much must go along with the flow of events surrounding us, which when working in insecure places often provokes anxiety and, when working among people in seemingly hopeless situations, impacts our emotional well-being. None of the participants in this study suggest that this can, or should be, avoided. On the contrary, they have used their sometimes hard-earned personal experiences to reflect creatively: on what can be done to better prepare younger colleagues for their first fieldwork and protect more experienced researchers from compassion fatigue or burnout; on how to enable institutional structures to provide better practices; and on how not to, themselves, contribute further to the co-creation of "no-go zones", or the nourishing of an "Indiana Jones" culture.

It is surely essential to acknowledge emotion as inherent in any research practice. But we must also promote researcher vulnerability as a valuable resource and not something to be prevented or avoided. One way to institutionally support this challenge would be to understand emotional counselling as a natural component of challenging research circumstances, rather than as an offer of treatment to those who return with traumatic symptoms. Potentially emotionally challenging situations should therefore be included in project design and methodology and form part of the evaluation made by ethics boards and the supervision given to younger scholars by their more experienced senior colleagues. One way to facilitate such institutional change, several younger colleagues have reminded me, is the power of the example: experienced researchers could certainly be more open about their own past and present vulnerabilities. For a start we could let go of the Indiana Jones culture and collectively search for more respectful, empathetic—and in the end more effective—ways of building trust with our interlocutors.

Time is a precious resource, especially when seeking to generate as much knowledge as possible while in the field. Participants in this study nevertheless pointed to the need for relaxation and leisure activities when conducting fieldwork over long periods of time. They also stressed the importance of maintaining research diaries, both as a cathartic tool for recording personal fears and shortcomings (Browne 2013), but also for expressing personal feelings of guilt, apprehension, and worry. Apart from serving as a logging device or a mere self-investigatory outlet, field diaries can then be used more explicitly to examine the ways our personal challenges and emotions impact the research process and its outcomes (Punch 2012). Those who have kept such diaries have moreover found them extremely helpful when later experimenting with more dialogical or creative forms of writing.

The attitude to sometimes dreaded ethical review boards and mandatory debriefing after completed fieldwork was more divided. With regard to the first, several have experienced that it is institutional interest rather than employee well-being that is protected. About debriefing, some find that in the best of cases it makes little difference, in the worst of cases it leads to retraumatisation. Others have found available debriefing structures helpful, but note that they ought not to be standardised, nor reflect institutional power structures. Few wish to involve individuals with power over their present and future work opportunities (directors, supervisors, human resource personnel) in such structures, but see potential in creating "safe groups" that meet regularly, can be contacted during fieldwork, and always when someone comes back from the field.

Interestingly, instances of compassion fatigue seem to have as much to do with the prevailing migration policy discourse in the Global North as with overexposure to misery and uphill struggles encountered in the field. This suggests that "compassion fatigue" is not necessarily a sign of a lack of compassion, but rather—and on top of the current emotional politics in research environments—is reflective of the politicised context in which contemporary migration research is carried out.

Stefanie Kappler (2021) reminds us that solidarity is a powerful response to the power differentials that emerge from different levels of privilege. Privilege not only determines mobility—who has permission to move in and out of which spaces—but also what researchers are willing and capable of observing and writing about, and for how long. This obviously makes it crucial to avoid making empty promises to the people we engage with, but perhaps also to be realistic about how much we can do. Ultimately, it offers an opportunity to rethink and push for more ethical—and more equal—research practices, or, as suggested by Abedi et al. (2020), new, non-exploitative forms of knowledge production.

NOTE

1. For a more explicit focus on vicarious trauma among researchers, I refer to Appelbaum (2008), Coles et al. (2014), Plakas (2018), Plüg and Collins (2018) and Sloan et al. (2019).

BIBLIOGRAPHY

Abedi Dunia, Oscar, Stanislas Bisimwa, Elisée Cirhuza, Maria Eriksson Baaz, John Ferekani, Pascal Imili, Evariste Kambale et al. 2019. "Moving Out of the Backstage: How Can We Decolonialize Research?" Accessed January 7, 2021. https://thedisorderofthings.com/2019/10/22/moving-out-of-the-backstage-how-can-we-decolonize-research/.

Abedi, Oscar, Maria Eriksson Baaz, David Mwambari, Swati Parashar, Anju Oseema Maria Toppo, and James Vincent. 2020. "The Covid-19 Opportunity: Creating More Ethical and Sustainable Research Practices." *Items, Insights from the Social Sciences*. Accessed January 7, 2021. https://items.ssrc.org/covid-19-and-the-social-sciences/social-research-and-insecurity/the-covid-19-opportunity-creating-more-ethical-and-sustainable-research-practices/.

Åhäll, Linda. 2018. "Affect as Methodology: Feminism and the Politics of Emotion." *International Political Sociology* 12 (1): 36–52.

Ahmed, Sara. 2014. *The Cultural Politics of Emotion*. 2nd ed. Edinburgh: Edinburgh University Press.

Andersson, Ruben. 2019. *No Go World: How Fear is Redrawing our Maps and Infecting our Politics*. Oakland: University of California Press.

Appelbaum, Jenna. 2008. "Trauma and Research: Bearing Responsibility and Witness." *Women's Studies Quarterly* 36 (1): 272–275.

Bosworth, Mary. 2017. "Secondary Trauma and Research." Accessed January 7, 2021. https://www.law.ox.ac.uk/research-subject-groups/centre-criminology/centreborder-criminologies/blog/2017/10/secondary-trauma.

Browne, Brendan Ciaran. 2013. "Recording the Personal: the Benefits in Maintaining Research Diaries for Documenting the Emotional and Practical

Challenges of Fieldwork in Unfamiliar Settings." *International Journal of Qualitative Methods* 12 (1): 420–435.

Carling, Jørgen. 2002. "Migration in the Age of Involuntary Immobility: Theoretical Reflections and Cape Verdian Experiences." *Journal of Ethnic and Migration Studies* 28 (1): 5–42.

Castellanos, M. Bianet. 2019. "Introduction." In *Detours: Travel and the Ethic of Research in the Global South*, edited by M. Bianet Castellanos, 3–18. Tucson, AZ: The University of Arizona Press.

Coles, Jan, Jill Astbury, Elizabeth Dartnall, and Shaznee Limjerwala. 2014. "A Qualitative Exploration of Researcher Trauma and Researchers' Response to Investigating Sexual Violence." *Violence Against Women* 20 (1): 95–117.

Distretti, Emilio. 2020. "Enforced Disappearance and Border Deaths Along the Migrant Trail." In *Border Deaths: Causes, Dynamics and Consequences of Migration-related Mortality*, edited by Paolo Cuttitta, and Tamara Last. Amsterdam: Amsterdam University Press.

Felbab-Brown, Vanda. 2014. "Security Considerations for Conducting Fieldwork in Highly Dangerous Places or on Highly Dangerous Subjects." Accessed January 7, 2021. http://webarchive.ssrc.org/working-papers/DSD_ResearchSecurity_03_Felbab-Brown.pdf.

Fleetwood, Jennifer. 2009. "Emotional Work: Ethnographic Fieldwork in Prisons in Ecuador." *eSharp, Special Issue: Critical Issues in Researching Hidden Communities*, 28–50. Accessed August 1, 2021. https://core.ac.uk/download/pdf/84339424.pdf.

Goldstein, Daniel M. 2014. "Qualitative Research in Dangerous Places: Becoming an Ethnographer of Violence and Personal Safety." Accessed January 7, 2021. http://webarchive.ssrc.org/working-papers/DSD_ResearchSecurity_01_Goldstein.pdf.

Grimm, Jannis J, Kevin Koehler, Ellen M. Lust, Ilyas Saliba, and Isabell Schierenbeck. 2020. *Safer Field Research in the Social Sciences*. London: Sage Publications.

Hoffmann, Kasper. 2014. *Caught between Apprehension and Comprehension: Dilemmas of Immersion in a Conflict Setting*. Copenhagen: DIIS Working Paper (No. 9).

Hugman, Richard, Eileen Pittaway, and Linda Bartolomei. 2011. "When 'Do No Harm' is Not Enough: the Ethics of Research with Refugees and Other Vulnerable Groups." *British Journal of Social Work* 41 (7): 1271–1287.

Kalinga, Chisomo. 2019. "Caught Between a Rock and a Hard Place: Navigating Global Research Partnerships in the Global South as an Indigenous Researcher." *Journal of African Cultural Studies* 31 (3): 270–272.

Kappler, Stefanie. 2021. "Privilege." In *The Companion to Peace and Conflict Fieldwork*, edited by Roger Mac Ginty, Roddy Brett, and Birte Vogel. London: Palgrave Macmillan.

Khosravi, Shahram. 2018. *After Deportation: Ethnographic Perspectives.* Basingstoke: Palgrave Macmillan.

Kozinets, Robert. 2015. *Netnography: Redefined.* London: Sage Publications.

Kristensen, Nanna Hauge. 2017. "Det Grænseoverskridende Feltarbejde." *Jordens Folk* 3–4 (52): 70–77.

Lake, Milli, and Sarah E. Parkinson. 2017. "The Ethics of Fieldwork Preparedness." Accessed January 7, 2021. https://politicalviolenceataglance.org/2017/06/05/the-ethics-of-fieldwork-preparedness/.

Last, Tamara. 2020. "Secondary Trauma Among Researchers: Implications of Traumatogenic Research in Archives." *SAGE Research Methods Cases.* Accessed January 7, 2021. https://doi.org/10.4135/9781529708462.

Lubkemann, Stephen C. 2008. "Involuntary Immobility: on a Theoretical Invisibility in Forced Migration Studies." *Journal of Refugee Studies* 21 (4): 454–475.

Mac Ginty, Roger, Roddy Brett, and Birte Vogel (eds.). 2021. *The Companion to Peace and Conflict Fieldwork.* London: Palgrave Macmillan.

Mackenzie, Catriona, Christopher McDowell, and Eileen Pittaway. 2007. "Beyond 'Do No Harm': The Challenge of Constructing Ethical Relationships in Refugee Research." *Journal of Refugee Studies* 20 (2): 299–319.

Marinovich, Greg, and João Silva. 2000. *The Bang-Bang Club: Snapshots from a Hidden War.* New York: Basic Books.

Martin-Ortega, Olga, and Johanna Helman. 2009. "There and Back: Surviving Research in Violent and Difficult Situations." In *Surviving Field Research: Working in Violent and Difficult Situations*, edited by Chandra Lekha Shiram, John C. King, Julie A. Mertus, Olga Martin-Ortega, and Johanna Herman. Abingdon, Oxon: Routledge.

McLean, Athena (with Annette Leibing). 2011. "Ethnography and Self-Exploration." *Medische Antropologie* 23 (1): 183–201.

Menjívar, Cecilia, Marie Ruiz, and Immanuel Ness. 2019. *The Oxford Handbook of Migration Crises.* New York: Oxford University Press.

Môcnik, Nena. 2020. "Rethinking Exposure to Trauma and Self-Care in Fieldwork-Based Social Research: Introduction to Special Issue." *Social Epistemology* 34 (1): 1–11.

Parvez, Z. Fareen. 2018. "The Sorrow of Parting: Ethnographic Depth and the Role of Emotions." *Journal of Contemporary Ethnography* 47 (4): 454–483.

Plakas, Christina. 2018. "Burnout, Compassion Fatigue, and Secondary Traumatic Stress Among Humanitarian Aid Workers in Jordan." Accessed January 7, 2021. https://www.researchgate.net/publication/328685237_Burnout_Compassion_Fatigue_and_Secondary_Traumatic_Stress_Among_Humanitarian_Aid_Workers_in_Jordan.

Plüg, Simóne, and Anthony Collins. 2018. "It Hurts to Help: Vicarious Trauma in Sensitive Research and Community Projects in South Africa." *The Australian Community Psychologist* 29 (1): 22–35.

Punch, Samantha. 2012. "Hidden Struggles of Fieldwork: Exploring the Role and Use of Field Diaries." *Emotion, Space and Society* 5 (2): 86–93.

Richmond, Anthony H. 1994. *Global Apartheid: Refugees, Racism and the New World Order*. Oxford: Oxford University Press.

Shah, Sonia. 2020. *The Next Great Migration: The Story of Movement on a Changing Planet*. London: Bloomsbury.

Shiram, Chandra Lekha, John C. King, Julie A. Mertus, Olga Martin-Ortega, and Johanna Herman (eds.). 2009. *Surviving Field Research: Working in Violent and Difficult Situations*. Abingdon, Oxon: Routledge.

Sloan, Katie, Jennifer Vanderfluit, and Jennifer Douglas. 2019. "Not 'just my problem to handle': Emerging Themes on Secondary Trauma and Archivists." *Journal of Contemporary Archival Studies* 6: 20.

Stepputat, Finn, and Ninna Nyberg Sørensen. 2014. "Sociology and Forced Migration." In *The Oxford Handbook of Refugee and Forced Migration Studies*, edited by Elena Fiddian-Qasmiyeh, Gil Loescher, Katy Long, and Nandy Sigona. Oxford: Oxford University Press.

Sørensen, Ninna Nyberg, and Laura Huttunen. 2020. "Missing Migrants and the Politics of Disappearance in Armed Conflicts and Migratory Contexts." *Ethnos*. https://doi.org/10.1080/00141844.2019.1697333.

Sørensen, Ninna Nyberg, and Sine Plambech. 2019. *Global Perspectives on Humanitarianism*. DIIS Report No. 03. Copenhagen: Danish Institute for International Studies.

Theidon, Kimberly. 2014. "How was your trip? Self-Care for Researchers Working And Writing on Violence." Accessed January 7, 2021. http://web archive.ssrc.org/working-papers/DSD_ResearchSecurity_02_Theidon.pdf.

Turner, Sarah. 2010. "Research Note: The Silenced Assistant. Reflections of Invisible Interpreters and Research Assistants." *Asia Pacific Viewpoint* 51 (2): 206–219.

Woon, Chih Yuan. 2013. "For 'Emotional Fieldwork' in Critical Geopolitical Research on Violence and Terrorism." *Political Geography* 33 (1): 31–41.

PART III

Comments

CHAPTER 9

On Data and Care in Migration Contexts

Koen Leurs

[O]ur current political and sociotechnical moment sits at the forefront of philosophical questions about who cares, how they do it, and for what reason.
—Hi'ilei Julia Kawehipuaakahaopulani Hobart and Tamara Kneese (2020, 2)

Offering fresh impetus to the emerging interdisciplinary research focus area of digital migration studies, this book probes how migration researchers, practitioners and policymakers can care for data. In taking a caring perspective, this book offers an important response to the recent trend of seeing migration as a laboratory where experiments with big data can be conducted. In particular, "irregularised migrants", the group of mobile subjects on which this anthology focuses, become guinea pigs. Experimental big-data-driven technosolutionism must be understood in a broader socio-political context where refugee and asylum migratory

K. Leurs (✉)
Department of Media and Culture, Utrecht University, Utrecht, The Netherlands
e-mail: K.H.A.Leurs@uu.nl

M. Sandberg et al. (eds.), *Research Methodologies and Ethical Challenges in Digital Migration Studies*, Approaches to Social Inequality and Difference, https://doi.org/10.1007/978-3-030-81226-3_9

movements are taken to stage a crisis, e.g. the so-called "European migration crisis". Rather than caring for, under the heading of crisis, the key aim is increasingly to control mobile groups through datafied solutions. For example, Frontex, Europe's Border and Coast Guard Agency is carrying out research and innovation based on big-data-driven artificial intelligence (AI) as part of its "Integrated Border Management" (2021, 62), which seeks to identify, contain and deter particular mobile people. Technologies to be tested include "automated border control", "small autonomous unmanned aerial systems" and "geospatial data analytics of operational awareness" (ibid., 35–40). Refugee camps in Greece such as "Moria 2.0" become the EU's "sandbox for surveillance technologies", where data-driven securitisation plans include "camera surveillance with motion analysis algorithms monitoring the behavior and movement of centre residents" (Molnar 2021). The "long summer of migration" in 2015, and additional so-called migration crises, with Rohingya in Bangladesh, Venezuelans in Brazil and South Sudanese fleeing to Kenya, illustrate the allure of what we can call the "big data sublime" (Mosco 2004): assuming technological innovation can disrupt and solve problems (without having to come to terms with underpinning large-scale historical, socio-cultural, geo-political and economic concerns). The UN Special Rapporteur on Racism, Racial Discrimination, Xenophobia and Related Intolerance sums up by commenting that governments and UN agencies "are subjecting refugees, migrants, stateless persons and others to human-rights violations, and extracting large quantities of data from them on exploitative terms that strip these groups of fundamental human agency and dignity" (Achiume 2020, 1). In thinking further how we as critical, engaged migration, border and media researchers operate in this space, in this commentary, I draw out conceptual assumptions around the two main thematics of data and care that underpin the contributions of this field-setting book.

DATA

As a philosophical term, data can be etymologically traced back to the seventeenth century Latin plural of "datum", meaning literally "something given" (Lexico 2021). The assumption of (big) data as a given objective, a fixed and factual neutral rendering of reality, remains a key rallying point in heated academic and activist discussions. The notions of dataism and data colonialism have recently begun to dominate these discussions. Commonly understood as mutually exclusive analytic lenses, these opposite sides of the debate can lure researchers and others into the trap of feeling they must choose one or the other. On the one hand, the notion of dataism captures an ideological "belief in objective quantification" and "trust in the (institutional) agents" that gather, analyse and share data (Van Dijck 2014, 198; Harari 2017). From this perspective, data and algorithms gain increasing authority over decision-making practices. Ideologies of dataism are also apparent in some domains of academia, including migration studies, as well as migration management, policymaking and humanitarianism, as several chapters of this book demonstrate. For example, in Chapter 5, Laura Stielike convincingly critiques the exploitative machinic vision discernible in big-data-based knowledge production on migration. The assumption that "migration as something that needs to be governed and that can be better governed through better data" reflects the digital sublime aspect of dataism and illustrates how certain strands of migration research reinforce the assumption that contemporary migration is inherently tied to database-driven governmentality.

The concept of "data colonialism" has grown into a prominent alternative critical framework. Jim Thatcher, David O'Sullivan and Dillon Mahmoudi proposed the term to dismantle utopian imaginaries of "digital frontierism" and grasp "accumulation by dispossession" (2016, 990). In a similar vein, Nick Couldry and Ulises A. Mejias draw on the term to address datafication as a new expansive formation of capitalism: "data relations enact a new form of data colonialism, normalizing the exploitation of human beings through data" (2019, 336). There is a growing body of work critiquing experimental governmentality projects that colonise datafied migrant populations under the heading of crisis management, securitisation and risk mitigation or "data-driven humanitarianism" (e.g. Molnar 2019). The book also delves further into the oppressive and exploitative data relations observable in the field of migration.

To account for the multiplicity, heterogeneity and inherent situatedness of data, datafication and data practices, Stefania Milan and Emiliano Treré rightly invite scholars to move beyond data universalisms such as dataism and data colonialism (2019). They argue that "the main problem with data universalism is that it is asocial and ahistorical, presenting technology (and datafication-related dynamics, we add) as something operating outside of history and of specific sociopolitical, cultural, and economic contexts" (2019, 324) and call for epistemic diversity and most importantly epistemic justice. Stefano Calzati similarly argues that to operationalise "pluralisation" of our understandings of data, we should speak about dataisms and data colonialisms (2020, 4).

This book proposes various ways to address data as inherently situated, contextualised and partial forms of representation and narratives. To draw out this understanding further, we can take cues from established theories on the performativity of language and performativity of images. Following J. L. Austin's book *How to do things with words?* (1962) we can draw on speech-act theory to conceptually pursue the question *How to do things with data?* In addition, we can address how data stretches across narrative and visual domains by taking cues from W. J. T. Mitchell's book *What do pictures want?* to ask *What is said about data?* And *what does data want?* For this purpose, in the remainder of this section, I will take the following 4 steps, (I) discuss some of the theoretical premises of performativity, (II) address performativity vis-à-vis power relations, (III) explicate how these processes are constituted through digital networks and datafication and (IV) transpose these insights into the specific context of datafied migration.

I. For the language philosopher John L. Austin, particular types of discourse can be understood as illocutionary acts. Illocutionary acts are performative utterances such as, for example, "You're fired", "I apologise", "You are under arrest" or "I now pronounce you married". Depending on the person making these statements and the contexts in which they are made, these speech acts result in a changed state of affairs or change in our relationships and social world: "The uttering of a performative is, or is part of, the doing of a certain kind of action, the performance of which, again, would not normally be described as just 'saying' or 'describing' something" (Austin 1962, 5). Besides Austin's language philosophy, theories of performativity have been developed in phenomenology and existentialism, ethnomethodology as well as theatre and performance studies (Isin 2021).

II. Theories of power in relation to performativity have been developed in queer and black feminism, poststructuralism, decolonial and postcolonial studies, and science and technology studies (Isin 2021). Articulating the relationship between discourse, gender normativity, sexuality and racism, Judith Butler conceptualised how performativity is inherently power-ridden. With her notion of the performative, Butler goes beyond distinctions between material-embodied and symbolic-discursive domains. Gender performativity is the constitutive stylised repetitious process through which one acquires a gendered subjectivity: "language sustains the body not by bringing it into being or feeding it in a literal way; rather, it is by being interpellated within the terms of language that a certain social existence of the body first becomes possible" (Butler 1997, 6). Performativity establishes and reinforces power relations through "the repetition or citation of a prior, authoritative set of practices" (Butler 1993, 226). However, theories of performativity seek to create awareness of how room manoeuvre, contestation and agency always remain. For Butler, gender can, for example, be subverted by unsettling and denaturalising it as "an act [...] which is open to splitting, self-parody, self-criticism" (1990, 282).

III. In recent years, several scholars have taken up performativity as a critical lens to address big data and datafication. Taking the example of databased search procedures, Rita Raley understands datafied surveillance (dataveillance) as a performative process: "Our data bodies [...] are repeatedly enacted. Data is in this respect *performative*: the composition of flecks and bits of data into a profile" (2013, 218). Larissa Hjorth illustrates how families do digital kinship, they also take up dataveillance as an empowering form of agency, to find reassurance in being able to monitor and share mundane performative activities over distance (2021). Focussing on performativity as agential, Engin Isin and Evelyn Ruppert have theorised digital citizenship as a form of politics in their reflection on "how people perform themselves as political subjects by making digital rights claims [...] by saying and doing something through the internet" (2020, xi).

IV. Drawing on Butler's notion of citationality, Tobias Matzner specifies how normativity shapes the performative big-data forms of migration governance (2016). For Matzner, through its diagnoses, or "verdicts", big data develop a "subjectivising force", for example, in the context of border control or processing mobile populations. Through that, they install hierarchy as they "cite a particular norm to count as a subject in the first place": a subject who is allowed to pass or needs to be stopped at a border, a subject allowed or denied a visa or allowed to board a plane (2016, 206–207). Vassilis S. Tsianos and Brigitta Kuster have called for critical scrutiny of datafied migration, while emphasising "doing digital borders", which emphasises the digital "performativity of the border" (2016, 237). From a separate angle, Stephan Scheel, Evelyn Ruppert and Funda Ustek-Spilda address the increasing data-driven knowledge making practices of migration across policy and academic domains into performative enactments (2019). As the processual understanding of "migrants' digital space" by Vasiliki Makrygianni, Ahmad Kamal, Luca Rossi and Vasilis Galis (Chapter 2) illustrates, the present book adopts a constructivist view of the performativity of data practices in migration. The book shows this perspective has the potential to reveal the intricate interplay between how the oppressive workings of technological systems, which variously play out, are contested and negotiated in the everyday lived experiences of distinctively situated mobile groups.

CARING FOR (BIG) DATA

After having articulated further alternative, nuanced and holistic understandings of data as performative, we now turn to the second key term underpinning the book: care. Marie Sandberg and Luca Rossi propose that for digital migration researchers "approaching migrants' digital data collection 'with care' means pursuing a more critical approach to the use of big data in migration research where the data is not an unquestionable proxy for social activity" (in the introductory chapter for this book). Here, they base their productive thoughts on Annemarie Mol's understanding of "care in practice" as a heterogeneous process resulting from a myriad of socio-material actors interacting in specific conditional settings (Mol et al. 2010). Sandberg and Rossi call for a theory and ethics of careful data research in and on migration, and below I address some genealogies to further situate the potentialities of pursuing caring migration data relations.

From an etymological perspective, we can trace caring to the Old English verb of "carian", which is of Germanic origin; and can be related to Old High German words "chara", meaning "grief, lament" and the Old Norse word "kǫr", meaning "sickbed" (Lexico 2021). Commonsensical understandings of care revolve around forms of provision reflecting what is needed for the health, welfare, maintenance and protection of something or someone; as well as attention applied to doing something right and avoiding damage or risk (Lexico 2021). Over time, these various meanings have lingered variously in care projects, and have exacerbated and deepened inequalities and hierarchies. Daniela Agostinho recognises the potential of care as a generative framework for critical "thinking about life and livability under digital and datafied conditions", but quite rightly cautions that "the racial, gendered, and colonial histories of care make it a difficult concept to think and work with" (2021, 80–81).

Historically, care is a term strongly reconceptualised by feminist theorists, elevating it from its previous marginalised feminine rendering of care duties to a lens for scrutinising academic knowledge production practices. A ground-breaking feminist definition of care was proposed by Joan Tronto and Bernice Fisher, who prioritised care as a material moral foundation to serve reproduction of socially just human life:

> On the most general level, we suggest that caring be viewed as a species activity that includes everything that we do to maintain, continue, and repair our 'world' so that we can live in it as well as possible. That world includes our bodies, our selves, and our environment all of which we seek to interweave in a complex, life-sustaining web. (1990, 40)

Like the practice-oriented understanding of data discussed above, caring can first be seen as an active practice revolving around the basic aims of survival, sustaining capabilities and avoiding suffering. Ethics of care is further premised on the moral dimensions of attentiveness, responsibility, competence and responsiveness (Tronto and Fisher 1990). Caring is often perceived as dominantly other-oriented, but self-care should also be recognised as a legitimate aim, and it functions as a prerequisite for providing care to others (Engster 2005). When informing research practice, a feminist ethics of care potentially provides a paradigm change, in offering alternative guidelines for conducting ethical human-centric research.

Feminist sociologists Rosalind Edwards and Melanie Mauthner offer a productive set of guiding questions that can assist researchers in translating feminist ethics of care ideals into their research practice:

- Who are the people involved in and affected by the ethical dilemma raised in the research?
- What is the context of the dilemma in terms of the specific topic of the research and the issues it raises personally and socially for those involved?
- What are the specific social and personal locations of the people involved in relation to each other?
- What are the needs of those involved and how are they interrelated?
- Who am I identifying with, who am I posing as otherwise, and why?

- What is the balance of personal and social power between those involved?
- How will those involved understand our actions and are these in balance with our judgement about our own practices?
- How can we best communicate the ethical dilemmas to those involved, give them room to raise their views, and negotiate with and between them?
- How will our actions affect relationships between the people involved? (2002, 28–29)

These questions are generative to pursue reflecting on the ambiguous "tinkering" practices that for Mol, Moser and Pols constitute care relations with all technologies (2010).

These questions invite researchers to engage in a reflexive mode of knowledge production on digital migration. Ethics of care concerns include obstacles, refusals as well as positionalities of researchers and research communities, which are often not explicitly addressed in published research output or conference presentations. Elsewhere, colleagues and I have critiqued this process under the heading of "dirty methods" (Bivens, Harvey, Leurs, Luka, Milette, Shepherd, forthcoming) where we argue "students are expected to learn how to do research through imitation, largely as a form of individualized apprenticeship, and consult methodological textbooks that offer clean, rigid, and disembodied recipes but pay little attention to the management of personal, collaborative, and bodily experiences, sensations, and anxieties" (ibid). With the notion of dirty methods, we seek to make generative the careful messiness and tinkering that is inherent in all research encounters.

In this book, (Chapter 6) Leandros Fischer and Martin Bak Jørgensen, Vasilis Galis (Chapter 7) and Ninna Nyberg Sørensen (Chapter 8) either implicitly or explicitly draw on ethics of care principles to articulate their own concerns as they seek to grapple with doing digital data research on migration with migrants. For example, Galis as well as Fischer and Jørgensen reflect on the role of academics and how they may be implicated in the migration industry, and whether, in avoiding complicity, researchers are required to become militant researchers. Another pivotal discussion concerns becoming aware of the plural subject position of researchers and the parallel need to attend to the plurality of informants' lives. Galis and Nyberg Sørensen highlight the complex emotional intensities of fieldwork for the various different types of bodies involved. An additional important reflexive step to be considered here could be to embrace all research perspectives as "partial perspectives" (Haraway 1988, 575). Following the thoughts of Donna Haraway, there is an inherent partiality in all research endeavours, therefore an ethical step could be to account for how decision-making and explicitly specifying how knowledge produced has come from somewhere, is being shaped by specific personal, emotional and epistemological trajectories, gatekeeping procedures and standpoints (Haraway 1988). To illustrate the importance of fore-grounding partiality, Fischer and Bak Jørgensen (Chapter 6) propose to operationalise caring digital research with mobile populations by drawing on Maurice Stierl, who provocatively argues that doing engaged research with migrants should amplify the epistemic crisis of European governmen-tality. For Stierl, "*Do harm* could be the motto for a critical and impactful scholarship of migration" (2020, 16). Fisher and Bak Jørgensen argue that in a politicised landscape, it is important to take explicit sides in social justice research, as not taking sides under the heading of objectivity may lead to the further silencing of already marginalised voices. By not taking sides, researchers are therefore implicated in the migration industry, contributing to an exclusionary and often violent, oppressive structure.

Further cues to reflect about and engage in caring relations with data may be gleaned from emerging discussions in feminist data studies and critical migration studies. In critical migration studies, Stephan Scheel and his colleagues reflect on care as a form of ethical accountability. They reflect on practicing three modes of care developed as part of their collaborative research project "thinking with others", "tinkering with field notes" and "dissenting within" (2020, 522). They built on Puig de la Bellacasa's relational understanding of care in knowledge production, understanding care as an "ontological requirement of relational worlds" (2012, 199). This aspect of relationality also shapes Marie Sandberg, Nina Grønlykke Mollerup and Luca Rossi's scrutiny of narrating mobility and bordering through "thin" social media and "thick" ethnographic data (Chapter 3).

In feminist media and data studies, several scholars have further devel-oped such a situated, responsive and careful approach to, for example, social media data analysis (e.g. Luka and Millette 2018).

So how should performative approaches to data and ethics of care be juxtaposed to stimulate creative debate? To operationalise a caring approach to (big) data, let us start with Catherine D'Ignazio and Lauren

F. Klein's seven data feminism principles: (I) Examine power, (II) Challenge power, (III) Elevate emotion and embodiment, (IV) Rethink binaries and hierarchies, (V) Embrace pluralism, (VI) Consider context and (VII) Make labour visible (2020, 17–18). A commitment to power, reflexivity and situatedness underpins these principles. Margie Cheesman's research on the role and imaginaries of blockchain in the refugee camps of Azraq and Zaatari in Jordan demonstrates the strong potential of such a situated, collaborative and careful research encounter with "irregularised migrants" on datafication. She grounds trust in blockchain technology by listening to refugee women and learning how they interpret their experiences through faith-based concepts. Research informants contest the datafication and digitisation of aid provision using the Islamic concept of "barakah" بَرَكَة to address the felt temporal inconsistency, immateriality and dependability of digital as opposed to cash assistance (Cheesman 2021). Future digital migration studies research can build on additional recently developed more situated, reflexive and affective forms of data studies, for example, taking the form of "data diaries" (Tkacz et al. 2021), the data "walkthrough method" (Light, Burgess and Duguay 2018), smartphone and social-media scroll-back methodologies (Georgiou and Leurs, Forthcoming 2021) and "data walking" (Van Es and de Lange 2020).

Conclusions

Migration, irregularised migration in particular, is shaped by a variety of human and non-human actors that sustain either care, securitisation or both. Care and securitisation are increasingly outsourced through automation, datafication and machinic vision. This volume breaks new ground by offering the means to reflect on the role of researchers at this conjuncture. In the present comment, I offer several genealogies of data and care, key themes that underpin the theorisation, methodological and ethical operationalisation of a more caring digital migration studies described in this book. Overall, the essays included urge us to go against the grain and rethink care-datafication technologies not as binary oppositions but as fundamentally relational. Also, in specifically situated contexts of datafication of irregularised migration, care is not to be seen as the absolute opposite of data. Rather, as the contributors show, pursuing careful data engagement allows understanding and reflection on the "ambivalence and shifting tensions" inherent in care-technology relations (Mol et al. 2010, 14). The genealogies discussed in the chapter remind us that neither migration nor data nor care are singular totalities. Social justice-oriented research on migration demands reflexive, situated and engaged careful tinkering with informants across scales and with human and non-human actors, to begin to come to terms with the paradoxes of data performativity. Data sets promise neutrality and total knowledgeability, but they are fundamentally ambiguous, power-ridden and "uncertain archives" (Bonde Thylstrup et al. 2021), with potentially humanising, dehumanising and abusive consequences for migrant subjects. Further scrutiny is required concerning the implications of the inherent uncertainty of data sets, of missing data, of the datafication of affect, and the politics of inclusion and exclusion that data encodes and challenges. In order to research datafied migration differently and sustain a caring, human-centric, profound and critical focus in the area of digital migration studies, we must encourage interdisciplinary dialogue between critical border and migration studies, social media studies, anthropology of migration, science and technology studies, critical data studies, feminist, queer and anti-racist theory.

BIBLIOGRAPHY

Achiume, E. Tendayie. 2020. "Report of the Special Rapporteur on Contemporary Forms of Racism, Racial Discrimination, Xenophobia and Related Intolerance." United Nations. Office of the High Commissioner, A/75/590. Retrieved from: https://www.ohchr.org/EN/newyork/Documents/A-75-590-AUV.docx.

Agostinho, Daniela. 2021. "Care." In *Uncertain Archives: Critical Keywords for Big Data*, edited by N. B. Thylstrup et al., 75–86. Cambridge, MA: MIT Press.

Austin, John L. 1962. *How to Do Things with Words*. Oxford: Oxford University Press.

Bivens, R., A. Harvey, K. Leurs, M. E. Luka, M. Milette, and T. Shepherd. Forthcoming. *Dirty Methods*. Waterloo, ON: Wilfred Laurier Press.

Butler, Judith. 1993. *Bodies That Matter: On the Discursive Limits of "Sex."* New York, NY: Routledge.

Butler, Judith. 1997. *Excitable Speech: A Politics of the Performative*. New York, NY: Routledge.

Butler, Judith. 1990. *Gender Trouble: Feminism and the Subversion of Identity.* New York, NY: Routledge.

Calzati, Stefano. 2020. "Decolonising 'Data Colonialism' Propositions for Investigating the Realpolitik of Today's Networked Ecology." *Television & New Media.* Online first https://doi.org/10.1177/1527476420957267.

Cheesman, Margie. 2021. "Reconceptualising Blockchain in Aid: From Trust in Institutions to Faith in Infrastructure." Paper presented at *Digital Practices and the Everyday conference,* panel 'Southern' responses to digital tech in (forced) migration. Utrecht University, the Netherlands, April 22, 2021.

Couldry, Nick, and Ulises A. Mejias. 2019. "Data Colonialism: Rethinking Big Data's Relation to the Contemporary Subject." *Television & New Media* 20 (4): 336–349.

D'ignazio, Catherine, and Lauren F. Klein. 2020. *Data Feminism.* Cambridge, MA: MIT Press.

Edwards, Rosalind, and Melanie Mauthner. 2002. "Ethics and Feminist Research: Theory and Practice." In *Ethics in Qualitative Research,* edited by M. Mauthner, M. Birch, J. Jessop, and T. Miller, 14–31. London: Sage.

Engster, Daniel. 2005. "Rethinking Care Theory: The Practice of Caring and the Obligation to Care." *Hypatia* 20: 50–74.

Frontex. 2021. "Artificial Intelligence-Based Capabilities for the European Border and Coast Guard." Retrieved from: https://frontex.europa.eu/media-centre/news/news-release/artificial-intelligence-based-capabilities-for-european-border-and-coast-guard-1Dczge.

Georgiou, M., and Koen Leurs. Forthcoming 2021. "Smartphones as Personal Digital Archives? Recentering Migrant Authority as Curating and Storytelling Subjects." *Journalism: Theory, Practice & Criticism,* special issue *Flesh Witnessing: Smartphones, UGC and Embodiment,* edited by Lilie Chouliaraki, and Mette Mortensen.

Harari, Yuval N. 2017. *Homo Deus: A Brief History of Tomorrow.* UK: Vintage Penguin Random House.

Haraway, Donna. 1988. "Situated Knowledges: The Science Question in Feminism and the Privilege of Partial Perspective." *Feminist Studies* 14 (3): 575–599.

Hjorth, Larissa. 2021. "Digital Kinship—Understanding Familial Care at a Distance: Keynote Migrant Belongings. Digital Practices and the Everyday Conference." Utrecht University, the Netherlands, 22 April 2021. Retrieved from https://www.youtube.com/watch?v=1YKdbYVKTd0.

Isin, Engin. 2021. "Digital Citizens Yet to Come." *Digital Practices and the Everyday conference.* Utrecht University, the Netherlands, 22 April 2021. Retrieved from https://www.youtube.com/watch?v=WGyTEZ5GpGY.

Isin, Engin, and Evelyn Ruppert. 2020. *Being Digital Citizens.* 2nd ed. London: Rowman & Littlefield.

Kawehipuaakahaopulani Hobart, H. J., and T. Kneese. (2020). Radical Care: Survival Strategies for Uncertain Times. *Social Text* 38 (1): 1–16.

Lexico. 2021. "UK Dictionary". Oxford University Press. Retrieved from: https://www.lexico.com/definition/.

Light, Ben, Jena Burgess, and Stefanie Duguay. 2018. "The Walkthrough Method: An Approach to the Study of Apps." *New Media & Society* 20 (3): 881–900.

Luka, Mary E., and Mélanie Millette. 2018. "(Re)Framing Big Data: Activating Situated Knowledges and a Feminist Ethics of Care in Social Media Research." *Social Media + Society*. https://doi.org/10.1177/2056305118768297.

Matzner, Tobias. 2016. "Beyond Data as Representation: The Performativity of Big Data in Surveillance." *Surveillance & Society* 14 (2): 197–210.

Milan, Stefania, and Emiliano Treré. 2019. "Big Data from the South(s): Beyond Data Universalism." *Television & New Media* 20 (4): 319–335.

Mol, Annemarie, Ingunn Moser, and Jeanettte Pols. 2010. "Care: Putting Practice into Theory." In *Care in Practice: On Tinkering in Clinics, Homes and Farms*, edited by A. Mol, I. Moser, and J. Pols, 7–26. Bielefeld: Transcript Verlag.

Molnar, Petra. 2019. "New Technologies in Migration: Human Rights Impacts." *Forced Migration Review*, June 7–9.

Molnar, Petra. 2021. "Moria 2.0: The Eu's Sandbox for Surveillance Technologies." *Euractiv*. Retrieved from: https://www.euractiv.com/section/digital/news/moria-2-0-the-eus-sandbox-for-surveillance-technologies/.

Mosco, Vincent. 2004. *The Digital Sublime*. Cambridge, MA: MIT Press.

Puig de la Bellacasa, Maria. 2012. "'Nothing Comes Without Its World': Thinking with Care." *The Sociological Review* 60 (2): 197–216.

Raley, Rita. 2013. "Dataveillance and Counterveillance." In *Raw Data Is an Oxymoron*, edited by Lisa Gitelman, 121–146. Cambridge, MA: MIT Press.

Scheel, Stephan, Francisca Grommeé, Evelyn Ruppert, F. Ustek-Spilda, B. Cakici, and V. Takala. 2020. Doing a Transversal Method: Developing an Ethics of Care in a Collaborative Research Project. *Global Networks* 20: 522–543.

Scheel, S., E. Ruppert, and Funda Ustek-Spilda. 2019. "Enacting Migration Through Data Practices." *Environment and Planning D: Society and Space* 37 (4): 579–588.

Stierl, Maurice. 2020. "Do No Harm? The Impact of Policy on Migration Scholarship." *Environment and Planning C: Politics and Space*, 2399654420965567.

Thatcher, Jim, David O'Sullivan, and Dillon Mahmoudi. 2016. Data Colonialism through Accumulation by Dispossession: New Metaphors for Daily Data. *Environment and Planning D: Society and Space* 34 (6): 990–1006.

Thylstrup, Bonde et al. 2021. "Big Data as Uncertain Archives." In *Uncertain Archives: Critical Keywords for Big Data*, edited by Nanna B. Thylstrup et al., 1–28. Cambridge, MA: MIT press.

Tkacz, Nathaniel, Mário da Mata Martins, Joao de Albuquerque, Flavio Horita, and D. Givanni Neto. 2021. "Data Diaries: A Situated Approach to the Study of Data." *Big Data & Society*. https://doi.org/10.1177/205395172 1996036.

Tronto, Joan C., and Berenice Fisher. 1990. "Toward a Feminist Theory of Caring." In *Circles of Care*, edited by E. Abel and M. Nelson, 36–54. SUNY Press.

Tsianos, Vasilis S., and B. Brigitte Kuster. 2016. "Eurodac in Times of Bigness: The Power of Big Data Within the Emerging European IT Agency." *Journal of Borderlands Studies* 31 (2): 235–249.

Van Dijck, José. 2014. "Datafication, Dataism and Dataveillance: Big Data between Scientific Paradigm and Ideology." *Surveillance & Society* 12 (2): 197–208.

Van Es, Karin, and Michiel de Lange. 2020. "Data with Its Boots on the Ground: Datawalking as Research Method." *European Journal of Communication* 35 (3): 278–289.

CHAPTER 10

Caring as Critical Proximity: A Call for Toolmaking in Digital Migration Studies

Anders Munk

The field of digital methods has emerged over the past two decades at the interface between media studies, science and technology studies (STS), computer science, and information design (Marres 2017; Rogers 2013). The core questions and ambitions covered closely resemble the kind of digital migration studies presented in this book. First, a commitment to studying wider sociocultural phenomena on, through, and with digital traces from online media. Second, a desire to critically engage with the role of these media, their platforms, and algorithms as infrastructures of social life. Third, a curiosity about how to wrongfoot entrenched divides between offline and online, actual and virtual, or qualitative and quantitative. And finally, ongoing experimentation with computational techniques for large-scale data capture analysis in conjunction with ethnographic approaches to deep sensemaking and interpretation. Indeed, to begin

A. Munk (✉)
University of Aalborg in Copenhagen, Copenhagen SV, Denmark
e-mail: anderskm@hum.aau.dk

© The Author(s) 2022
M. Sandberg et al. (eds.), *Research Methodologies and Ethical Challenges in Digital Migration Studies*, Approaches to Social Inequality and Difference, https://doi.org/10.1007/978-3-030-81226-3_10

235

caring for big data, as argued by Sandberg and Rossi in the introductory chapter to this book, is not only a welcome call for digital migration research but also rings true in digital methods more broadly.

The fact is that digital migration studies were early and significant contributors to the evolution of digital methods, although this may be relatively unacknowledged in both fields today. When Dana Diminescu began developing her E-Diasporas project in 2006,[1] she also began collaborating with a young research engineer, Mathieu Jacomy, who would later become one of the key toolmakers in digital methods (the story is recounted in Jacomy's recent Ph.D. dissertation, Jacomy 2021). Together with other team members, and spurred by the ambition of the project to collect and map relations between websites from thirty different migrant communities online (Diminescu 2008), they devised the concepts and developed the prototypes for two iconic research tools, namely, Gephi[2] (Jacomy et al. 2014), which is today one of the world's most popular pieces of open-source software for visual network analysis, and the Navicrawler[3] (Diminescu et al. 2011), which was a predecessor for the current state-of-the-art tool for web corpus curation in digital methods (Jacomy et al. 2016).

Reading through the timely and thought-provoking contributions in this book, it seems to me that the E-Diasporas collaboration has an important story to tell. What Jacomy and Diminescu practiced together in the borderland between digital methods and digital migration studies was not only a form of caring for big data, but a very particular form of caring that took place in what Bruno Latour would call "critical proximity" (Latour 2005) with the technical circumstances that surround the production and analysis of such data. Contrary to a critically distant position, which cares for the consequences of new technological practices (in this case data scientific practices) that encroach on a field (in this case digital migration studies), caring as critical proximity undertakes to intervene with such practices, reimagining what they could do and redesigning them accordingly. When Sandberg, Mollerup, and Rossi discuss what it would take to develop a "contrapuntal" analysis of digital connectedness (Chapter 3), or when Makrygianni, Kamal, Rossi, and Galis (Chapter 2) engage with how to map the relationality of digital migrant space, they do so in a form of critical proximity with the Facebook API and its shifting affordances. Like other forms of care, not least those described by scholars such as Annemarie Mol, Ingunn Moser, or Jeanette Pols (Mol 2008; Mol et al. 2015), Sandberg and Rossi draw on in their introduction; caring for big data could thus be construed as a material practice in which data-intensive analysis is *done* differently. Crucially, this *doing* depends on the willingness and ability of researchers to engage practically with the affordances of digital tools and platforms. As the chapters in this book demonstrate, this is not to be mistaken for simply replacing digital migration studies with computational social science. Caring for big data in critical proximity must necessarily entail that data science methods are reconceived and experimentally reassembled from a position firmly inside the field in question.

Caring for Digital Traces and Relational Spaces

In digital methods, caring as critical proximity has resulted in a wide variety of homegrown tools for harvesting and analysing digital traces from social media platforms such as Facebook (Rieder 2013) or Twitter (Borra and Rieder 2014). At the time of the E-Diasporas project, however, the main interest was still centred on websites. Several tools had already been developed for collecting websites, the so-called web corpora, when Jacomy and Diminescu began their collaboration, but their specific interest in online migrant communities made it necessary to do data curation and analysis differently.

Web pages at the time were (and to a large extent still are) written in hypertext markup language (html), which allows a browser to know which parts of a page to display as headings, body text, links, menus, images, etc. It also allows a piece of software known as a *scraper* to selectively harvest parts of that information, for example, all the links, all the images, or all the text found under certain headings. It does so by "repurposing" (Rogers 2013) the tags in the markup. A hyperlink (or hypertext reference) is marked up with the tag "href = " followed by a web address. Thus, by identifying all the instances of a "href = " in the html code and collecting the accompanying web addresses, a scraper piggybacks on the functionality and conventions of html, as well as the choices made by the authors of a web page, to build a data set. It follows that scrapers cannot simply harvest any kind of digital trace, at least not with equal ease, but must operate within an affordance space that has been designed elsewhere and by others, which is why Noortje Marres points out that such digital methods are "distributed" (Marres 2012), in the sense that they are conditioned by and constituted from a heterogeneous set of actors that are often extraneous to the immediate situation of the researcher.

Digital traces, then, are not simply given as data but actively taken as what Johanna Drucker (2011) calls "capta"—that which is taken as opposed to that which is given, in order to signal that there is nothing naturally occurring about it—with due regard to the specific sociotechnical circumstances of their construction. Caring for these sociotechnical circumstances is of critical importance to digital methods. The most obvious capta—those that are most straightforwardly capturable—are not necessarily the most interesting or amenable for a given research purpose. In the case of the E-Diasporas project, which was interested in mapping the online presence of specific diasporas through their websites and their linking practices, the generic hyperlink was not in itself a relevant digital trace to follow. Websites send links to other websites for a variety of reasons and clearly not exclusively to signal belonging to the same diaspora. The generic hyperlink, however, was the straightforwardly capturable digital trace. At the time, all the available web crawlers that made link scraping manageable and approachable for researchers did not distinguish between different types of hyperlinks. For good reasons: there were no tags or conventions in html that could be easily repurposed to make such a distinction, and in any case, the crawlers had been built for other purposes.

The predominant web crawler in digital methods, for instance, was called the Issue Crawler and had been developed with a particular media studies interest in mind (Marres and Rogers 2005). A crawler (or a spider) is a version of a scraper that follows hyperlinks from page to page according to a set of rules and thus automates part of the data collection process. The Issue Crawler took as input a list of seed pages dedicated to a specific issue (e.g., the GMO debate, a new vaccine, or a contested infrastructure project) and used them as starting points from which it followed all hyperlinks to more pages at a set crawl depth (number of link steps from the seed). This would produce a data set of more pages dedicated to the same issue, but also of pages about tangential issues, as well as news or social media sites, and internet infrastructure such as add servers, trackers, search engines, or content management systems. Rather than viewing this wider entanglement of web pages, which come into view when you generically follow hyperlinks from a seed, as noise, the Issue Crawler catered for a now long-standing tradition in digital methods for studying the infrastructures of online media as an integral part of the issue (Marres 2015).

When the knowledge interest is different, however, as was the case with E-Diasporas, the existence of ready-to-use tools like the Issue Crawler presents a challenge. Because they make certain digital traces, in this case the generic hyperlink, even more capturable than they already were, they require an extra level of care on behalf of the researcher, who must not only be capable of critically appreciating how capture could be different, but also intervene directly in the design of tools that make other forms of capta possible. This was exactly the approach taken by Jacomy and Diminescu in E-Diasporas when they realised that they would need to be able to curate multiple web corpora with a data collection approach that was simultaneously capable of leveraging the power of crawling to find all links from thousands of pages on the same website but also qualitatively select which of the discovered websites to include in a corpus. Rather than including all websites at a set distance from the seed, they needed a crawler that would continuously prompt the researcher to qualitatively decide if links should be followed or not. The solution was a purpose-built application for Firefox, the Navicrawler (Diminescu et al. 2011), which combined the qualitative element of browsing web pages with the quantitative power of crawling. The Navicrawler asks the user to navigate to a website from which to begin building the corpus. The crawler element visits all the pages on the site in question, scrapes all hyperlinks, and provides a list of discovered sites for the user, who then decides which to include in the corpus and which to leave out. When a site is included in the corpus, the crawler automatically reiterates the process on that site. In the case of E-Diasporas, the result was 30 different web corpora representing 30 different online migrant communities, each built from a known website of that community and curated through a selective crawling process whereby only sites relating to the same community were included.

The E-Diasporas web corpora, then, were made possible through caring as critical proximity. Indeed, this is true not only for the collection of the websites, which depended on the development of the Navicrawler, but also for the exploration of the relational spaces emerging from their linking practices, which depended on the development of the Gephi software (Jacomy et al. 2014). Visually exploring patterns in how websites were linked required network analysis and, in particular, force-directed network layouts. A force-directed layout produces a visualisation where nodes (in this case websites) are placed close to each other if they share many of the same connections (in this case hyperlinks). Force-directed layouts are non-deterministic algorithms that begin from a random positioning of the nodes and introduce an energy model that pushes them apart if they are unconnected. The visualisation will be different each time you run the algorithm, and settings for the energy model can be changed, further changing the visualisation. The goal is to exploratively obtain visualisations that prompt curiosity and help the researcher generate questions about the relational space that can be pursued qualitatively. Or rather, that is *one* goal in the use of force-directed layouts and typically one that is associated with practices such as multi-sited ethnography (Munk and Ellern 2015) or actor-network theory (Venturini et al. 2019) in which the phenomenon in question is not presumed to exist in a pre-existing bounded space, and the ongoing construction of spatiality is a central object of study. This requires a tool that allows the user to adjust the energy settings for the force-directed layout and experiment with producing different network views. Gephi was born as an answer to that specific knowledge interest (Fig. 10.1).

Implications for Digital Migration Studies

It is perhaps unlikely that digital migration studies anno 2021 will have to care much about the curation of websites. As the chapters in this book demonstrate, interest has, as in most other fields, shifted to digital platforms, not least social media. The need for caring as critical proximity has never been greater, though. The dependence on platform APIs for harvesting data has, over the years, proven highly unstable, as endpoints have been changed or deprecated and access denied at short notice. Simply relying on tools that others have built to do digital methods research quickly lands you in situations where whole research projects become untenable from month to month as the infrastructure on which the tool is based changes or is entirely removed. We must therefore be agile enough to not only think about, but also to develop in practice, makeshift alternatives (Perriam et al. 2020; Venturini and Rogers 2019).

This comes on top of an even more central point about critical proximity, which has always been true regardless of API changes and deprecations, namely, that specific digital traces must be repurposed for specific knowledge interests. As the E-Diasporas example demonstrates, this repurposing is often a matter of not capturing the most straightforwardly capturable digital traces but working materially to make other

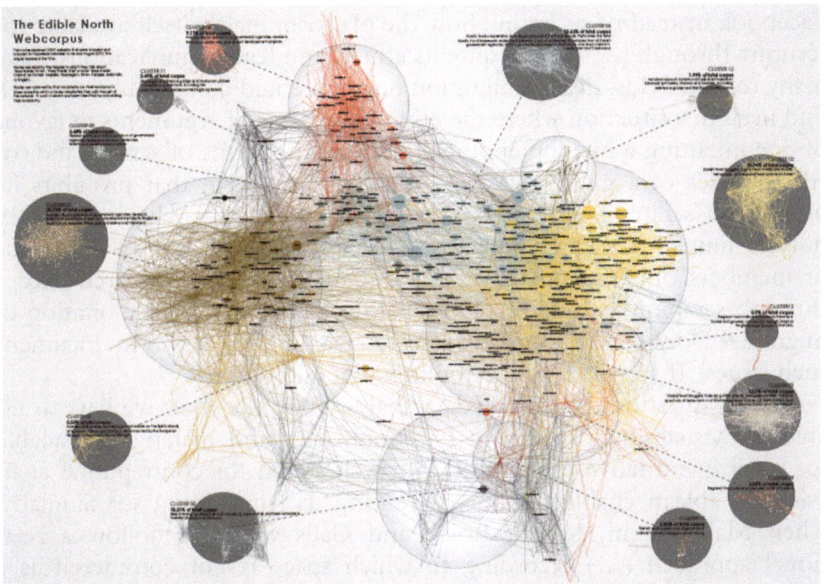

Fig. 10.1 An example of a web corpus collected with the Navicrawler and explored with a force-directed layout in Gephi (Munk 2019). By combining manual curation and automated crawling, I collected 2007 websites from food actors in Scandinavia. The relational space resulting from the linking practices of these actors becomes visible and explorable when the force-directed layout places clusters of interconnected websites in proximity with each other. The large yellow (right), red (top), and brown (left) clusters turn out to be national spheres of food-related websites (Danish, Norwegian, and Swedish), whereas some of the smaller clusters are communities of interest focussed on practices such as foraging, mushroom picking, beer brewing, or aquavit making. These relational spaces emerge from weblinks of the actors and become visible as such through the intervention of Gephi. This image is used with permission of the author [Rightsholder]

kinds of capta possible. When Sandberg, Mollerup, and Rossi convincingly argue for the merits of thinking about digital connectedness as "contrapuntal" (Chapter 3), then how should the Facebook API be repurposed to best support such an analysis? Is it even possible within the current restrictions on API-based research? If not, should we adopt a more activist stance (as suggested by Ben-David 2020) and begin scraping

Facebook instead of accepting how the platform makes itself available for scrutiny through the API, despite its ethical and legal complications? Like many related fields, digital migration research could conceivably very well find itself in a situation where the ethical and societal arguments in favour of documenting a phenomenon collide with the terms of service and/or API policies of a given platform. What if a platform that prohibits its users from scraping and offers no API access for research is also a key hub for human trafficking? What if a closed group with tens of thousands of members on Facebook, which is out of bounds for research except through scraping, is used for coordinating harassment or exploitation of migrants? Does digital migration research have an obligation to document such issues? If so, where is the toolbox that allows this?

Similar questions pertain to data analysis: Are the tools we have available for visualising patterns in Facebook data and making it available for quali-quantitative exploration also well suited for contrapuntal analysis? Is Tableau enabling us or preventing us from doing so? Similarly, when Makrygianni, Kamal, Rossi, and Galis want to "follow a relational approach (…) according to which space is not considered as a life container but as a derivative of social relations and interactions" (Chapter 2), are line graphs and custom error bar charts the visual techniques that really support that endeavour? And how would we go about collecting the relational user data today, now that the API no longer allows it?

These are questions for a research practice that cares for big data as a form of critical proximity. Care as a material practice obliges us to seriously consider how the sociotechnical circumstances of data capture and analysis can be engaged and transformed. This is as true in digital methods as it is in digital migration studies, and this book aptly demonstrates this point. It is not an easy task, however, as there will always be practical motivation to leave the data science to the data scientists and the question-posing to those in SSH with a digital research interest. It is simply convenient. But it is also dangerous and, I believe, the direct path to a version of caring for big data at a distance that will render our story of data science as either one of appropriation, where we accept remodelling our questions and research interests in order to become amenable to the methods of others, or one of antagonism, where we try to protect our qualitative and interpretive approaches by pointing out the biases, shortcomings, shallowness, and ethical problems of various kinds of big social data and computation. What this book promises to do instead—and it deserves a lot of credit for it—is to begin the hard work of reassembling data-intensive computational methods from a position within digital migration studies, in dialogue with the existing methodological landscape of the field, and in response to the research interests that are already articulated by its researchers.

NOTES

1. http://www.e-diasporas.fr/ [last accessed March 1, 2021].
2. https://gephi.org/ [last accessed March 1, 2021].
3. https://medialab.sciencespo.fr/en/tools/navicrawler/ [last accessed March 1, 2021].

BIBLIOGRAPHY

Ben-David, Anat. 2020. "Counter-Archiving Facebook." *European Journal of Communication*, 35 (3): 249–264. https://doi.org/10.1177/026732312 0922069.

Borra, Eris, and Bernhard Rieder. 2014. "Programmed Method: Developing a Toolset for Capturing and Analyzing Tweets." *Aslib Journal of Information Management*, 66 (3): 262–278. https://doi.org/10.1108/AJIM-09-2013-0094.

Diminescu, Dana. 2008. "The Connected Migrant: An Epistemological Manifesto." *Social Science Information*, 47 (4): 565–579.

Diminescu, Dana, Matthieu, Renault, Mehdi, Bourgeois and Jacomy, Mathieu. 2011. "Digital Diasporas Atlas Exploration and Cartography of Diasporas in Digital Networks." *Proceedings of the International AAAI Conference on Web and Social Media*, 5 (1).

Drucker, Johanna. 2011. "Humanities Approaches to Graphical Display." *Digital Humanities Quarterly*, 5 (1): 1–21.

Jacomy, Mathieu. 2021. *Situating Visual Network Analysis*. Aalborg University: Aalborg.

Jacomy, Mathieu, Paul Girard, Benjamin Ooghe, and Tomaso Venturini. 2016. "Hyphe, a Curation-Oriented Approach to Web Crawling for the Social Sciences." *International AAAI Conference on Web and Social Media*.

Jacomy, Mathieu, Venturini, Tomaso, Heymann, S., and Bastian, M. 2014. "ForceAtlas2, a Continuous Graph Layout Algorithm for Handy Network Visualization Designed for the Gephi Software." *PLoS One*, 9 (6): e98679.

Latour, Bruno. 2005. "Critical distance or critical proximity." Unpublished Manuscript. Accessed March 31, 2014. http://www.bruno-latour.fr/sites/def ault/files/p-113-haraway.pdf.

Marres, Nortje. 2012. "The Redistribution of Methods: On Intervention in Digital Social Research, Broadly Conceived." *The Sociological Review*, 60, 139–165.

Marres, Nortje. 2015. "Why Map Issues? On Controversy Analysis as a Digital Method." *Science, Technology, & Human Values*, 40 (5): 655–686. https://doi.org/10.1177/0162243915574602.

Marres, Nortje. 2017. *Digital Sociology: The Reinvention of Social Research*. John Polity & Sons: Cambridge.

Marres, Northe and Richars Rogers. 2005. *Recipe for Tracing the Fate of Issues and their Publics on the Web*.

Mol, Annemarie. 2008. *The Logic of Care: Health and the Problem of Patient Choice*. Abingdon: Routledge.

Mol, Annemarie., Ingunn Moser and Jeanette Pols. 2015. *Care in Practice: On Tinkering in Clinics, Homes and Farms* (Vol. 8). Transcript Verlag: Bielefeld.

Munk, Anders K. 2019. "Four Styles of Quali-Quantitative Analysis: Making Sense of the New Nordic Food Movement on the Web." *Nordicom Review*, 40 (s1): 159–176. https://doi.org/10.2478/nor-2019-0020.

Munk, Anders K., and Ellern, A. B. 2015. "Mapping the New Nordic Issuescape: How to Navigate a Diffuse Controversy with Digital Methods." *Tourism Encounters and Controversies: Ontological Politics of Tourism Development*, eds. R. Duim et al, Ashgate.

Perriam, Jessamy, Andreas Birkbak, and Andres Freeman. 2020. "Digital Methods in a Post-API Environment." *International Journal of Social Research Methodology*, 23 (3): 277–290.

Rieder, B. 2013. "Studying Facebook Via Data Extraction: The Netvizz Application." *Proceedings of the 5th Annual ACM Web Science Conference*, 346–355. https://doi.org/10.1145/2464464.2464475.

Rogers, Richard. 2013. *Digital Methods*. MIT Press.

Venturini, Tomaso, Anders K. Munk, and Mathieu Jacomy. 2019. "Actor-Network Versus Network Analysis Versus Digital Networks: Are We Talking about the Same Networks?" In *Digital STS: A Field Guide for Science & Technology Studies*. Princeton University Press.

Venturini, Tomaso and Richard Rogers. 2019. "'API-Based Research' or How can Digital Sociology and Journalism Studies Learn from the Facebook and Cambridge Analytical Data Breach." *Digital Journalism*, 7 (4): 532–540. https://doi.org/10.1080/21670811.2019.1591927.

What Should We Do as Intellectual Activists? A Comment on the Ethico-political in Knowledge Production

Anna Lundberg

"There must be those among whom we can sit down and weep and still be counted as warriors". This quote, by Adrienne Rich from her book *Your Native Land, Your Life* (1993), has haunted me numerous times while encountering children and their parents in voluntary legal advice groups—families who did not have their asylum claims recognised and therefore were irregularised. I encountered such families while providing counselling or conducting research together with self-organised groups in the Asylum Commission (Elsrud et al. 2021), conducting fieldwork at the Swedish Migration Agency, and working as an independent investigator for the Swedish government proposing legislative changes in the Aliens Act. Fluctuating between anger, hope, and hopelessness, at one time, I cross-stitched Rich's quotation on a pillow, as a gift for a doctoral student I supervised. During our sessions, she had often cried over the violence

A. Lundberg (✉)
Norrköping, Sweden
e-mail: anna.b.lundberg@liu.se

© The Author(s) 2022
M. Sandberg et al. (eds.), *Research Methodologies and Ethical Challenges in Digital Migration Studies*, Approaches to Social Inequality and Difference, https://doi.org/10.1007/978-3-030-81226-3_11

in the European migration control regime, the unjust refusal to permit the freedom of movement, and over her disappointment at not being able to translate the experiences she had gained during fieldwork into information that enabled her colleagues to understand the gravity of the current political situation.

Rich's "those among whom" is meaningful not only because it grants permission or even validates shedding tears, but also for encouraging collective efforts to develop epistemic communities of belonging (Yuval-Davis 2011), that is, spaces where we can converse about and enable transformative ethico-political research. By using the word ethico-political, I mean to invoke attentiveness, responsibility, curiosity, and the awareness that each one of us is capable of collective reflection (Bozalek and Zembylas 2017). It may also imply activism in Patricia Collin's sense of the word: politically engaged scholarship that may take place anywhere anytime, "because ideas and politics are everywhere" (Collins 2013, 37). Efforts to develop communities of belonging are also symbolised by a means of channelling frustration (and weeping) and persistent confrontation with the question *What should we do* as intellectual activists?

The classic ethical way to answer *What should we do?* is to either *enquire* what anyone who was similarly situated ought to do (*universalisation*), to ask how, within the scope of given resources, you can do your best for as many people as possible (*maximising*), or to develop *virtues*, such as generosity and truthfulness as a guide in decision making.

In this text, I will not advocate any of these ethical queries as the preferred way forward. Nor will I answer the question: *What should we do?* Instead, I want to speak in favour of creating spaces where we may develop as intellectual activists.

I will propose two concepts that are valuable for the creation of such spaces and that I, while reading the present book, see the authors referring to: epistemic injustice and hope. These concepts are also central to the work we do in the transformative collective initiative, the Asylum Commission (Elsrud et al. 2021). In collaboration between researchers, professionals, civil society actors, self-organised groups, and support networks, we examine and analyse the shifts and restrictions that have taken place in recent years' Swedish asylum regulations, as well as the consequences of these changes for people seeking protection, civil society, and welfare workers. The Commission's primary focus is a critical review and exchange of experiences, combined with continuous knowledge-dissemination work that is carried out based on asylum seekers' perspectives and lived experiences. Moreover, our ambition is also to think and act in new directions to enable a more solidaristic refugee policy in the future, attempting to provide what Fiorenza Picozza describes as "a contribution to an anticolonial political imagination that can sustain daily struggles against the asylum regime" (Picozza 2021, xxvi).

In the DIGINAUTS initiative, I see, in the same vein this endeavour to avoid harm, to be responsibly engaged, and to speak the truth to the powers that be (see Sandberg and Rossi, Chapter 1). The DIGIN-AUTS researchers, as seen in the contribution to this book, strive to provide spaces for exchanging experiences of political work with the unpredictability that such work always implies, whether you are in a mostly scientific or primarily activist milieu. I see an awareness that certain questions must be asked over and over again: How can this research benefit society without reproducing the state-sanctioned methods of violence called detention, deportation, rejection? What is "socially benefi-cial" research and what should it be in a field as politicised as migration in contemporary times of far-reaching and devastating neoliberal transforma-tions? Is it possible to be progressively exploratory in matters concerning the issue of how people seeking sanctuary can be welcomed, without simultaneously confirming the view that some people are not human enough to count as human, which underlies most political proposals in policy discussions about who deserves protection and (formal) inclusion?

In the chapter on emotional introspection, Ninna Nyberg Sørensen points to these queries in her view that "research results have little reso-nance among political decision makers and the general public, when the ranks of those 'for' and 'against' migration are more sharply drawn, and when humanitarian considerations become subordinated to wishes for control and restraint" (see Chapter 8). No matter how nuanced and committed we may be, and however valuable our research findings are for future policies, migration research that does not reproduce dominant images of migration and its notions of solutions tends to be silenced or emasculated. This is perhaps not so surprising, because, as Bimal Ghosh aptly observed two decades ago, "no other source of tension and anxiety has been more powerful [in the Global North] than the fear, both real and perceived, of huge waves of future emigration from poor and weak states in the years and decades to come" (Ghosh 2000, 10).

Bearing in mind these challenges and doubts concerning the contribu-tion of intellectual activism, in their chapter, Leandros Fischer and Martin

Bak Jørgensen highlight the expectations they faced within academia and when conducting fieldwork. The "complex responsibilities for high-quality and ethical research" are expectations that we need to meet (see Chapter 6). What impact does this have on the answer to a question that my colleagues and I are frequently asked when presenting the Asylum Commission's transformative work, namely: How do you distinguish between your role as a researcher and your role as an activist? How do you separate these roles? What do the expectations that Fischer and Jørgensen describe mean for the scepticism that migrants—unsurprisingly and well-justified as they are—convey when yet another researcher approaches them and makes claims on their time and stories (see Galis' discussion in Chapter 7). One argument put forward is that "there is a stronger imperative to publish less, better-quality research" (Fischer and Jørgensen, Chapter 6). Yet, we know that what counts as high-quality scholarship, what is deemed within its remit, and what it may accomplish in the social world remain very different, though this may be deeply unjust. Many articles in prestigious scientific journals are not openly available. If they are generally accessible, they are read-only by academics. They are rarely translated in a spirit of actively questioning the boundary between activism and academia or in contexts of mutual learning at the intersection of theory and practice. Open access in its current forms is not a solution here, because while it aims to combat commercial publishing channels, it simultaneously stimulates emerging platform monopolies.

Another aspect of engaged scholarship that emerges in this book covers the much needed problematisations of language and framing. In her chapter about big-data-based research, Laura Stielike investigates 17 peer-reviewed publications (Chapter 5). Stielike highlights how three mainstream migration narratives are produced in the papers, which boil down to a framing of migration as something that can be governed *more* (which is assumed to be necessary) through *better data* (which implies that research, in turn, is to provide the data). Much data-based migration research of today, Stielike concludes, reproduces an understanding of migration as a phenomenon that changes the size and composition of a population, and that can be influenced to a certain extent through political interventions, an assumed need to integrate migrants into receiving societies, and various forms of humanitarian assistance.

Within my research field, migration law, I also sense such more or less explicit agendas and assumptions, through the use of language, although this is often veiled by the supposedly impartial legal text. More rules are often requested in the name of the rule of law and legal certainty. Consider the concepts of economic migrants, third-country nationals, unaccompanied children, and voluntary return. How do we handle such legal-technical terms as they seep into academic literature? In our quest to influence, we easily unconsciously reproduce a policy or even a humanity that we do not think is good at all, or we realise fails on a multitude of important levels.

What to Do?

Giacomo Toffano and Kevin Smets (Chapter 4) present an inspiring pathway for new and transdisciplinary narratives in migration research in the Migration Trail. The trail is a visual representation of spatial data, initially assembled to accompany the narrative of two migrants on their journey to Europe. There is David, an ambitious Nigerian entrepreneur with big business plans that were a driving force in leading him through Libya, Italy, and France. And there is Sarah, a 19-year-old Syrian proceeding through the "Balkan Route" continually exchanging messages with her brother. The Migration Trail is an investigation of the interaction of textual, audio, visual, and spatial elements of communication drawing attention to the political, economic, and social dimensions of migration while, at the same time, reaffirming the humanity of people on the move. It represents something that is different from common press accounts and traditional data visualisations or fiction, where migrants often appear as mere victims: passive subjects of political–economic conditions in need of protection. The Migration Trail is indeed a protest against violent migration management and an example of how the repressive development can be criticised by intellectual activists.

All in all, the authors of this book demonstrate and provide alternate understandings of the dominant notions that permeate migration policies, in which people seeking refuge are depicted as "the other". The chapters reach beyond images of undesirable collective masses that must be kept away at all costs, or a vulnerabilised suffering individual who, if let in, will be admitted because it confirms "the us" as human beings (after all). This is done without ending up at the other extreme of the "curated success stories" (Fernandes 2017) where complex lived experiences are restructured into easily absorbable superficial stories. Even though such stories may be mobilised towards beneficial goals, the conditions under which they are told and the response to them involves a risk that the greater circumstances of global inequality will be concealed.

Epistemic injustice and hope are two concepts that can facilitate the future process of making sense of the question of "What should we do?" and supporting intellectual activists in their solidarity struggles.

EPISTEMIC INJUSTICE

Epistemic injustice was introduced by the moral philosopher Miranda Fricker in her book of the same name (*Epistemic Injustice. Power and the Ethics of Knowing*, 2007) to catch the interdependency between knowledge and power. Fricker explains how knowledge production is both political and ethical because it interacts with society's hierarchical arrangements, allowing some perspectives to come forward while others are left out. In the same vein, Gayatri Spivak (1988) has previously used the term "epistemic violence" to describe the disappearing of some knowledge, namely, that of marginalised groups, and how this undermines their ability to speak and be heard. And Pierre Bourdieu has insightfully stated that, "Among the most radical, surest, and best hidden censorships are those which exclude certain individuals from communication (e.g., by not inviting them to a place where people speak with authority, or by putting them in places without speech)" (Bourdieu 1977, 649).

The problem here is not just that others do not try to listen. Even though people in vulnerable positions might be asked to speak, they are rarely considered to be people who know and understand (Boochani et al. 2020). This negligence is seen in much migration research, where migrants are often portrayed as objects without being acknowledged as co-producers of academic knowledge (Grosfoguel et al. 2015).

According to Fricker, epistemic injustice is a form of "double structural discrimination" that is enabled because privileged groups have primary control over society's analytical resources. This discrimination is manifested through the questioning of certain persons' credibility as knowers, called testimonial injustice, and excluding collective interpretive resources that reproduce a form of hermeneutical injustice, i.e., injustice based on the theory and methodology of interpretation, especially of biblical texts. This occurs when the narrative and interpretive resources necessary to describe and understand the experiences of marginalised groups are lacking. It leaves the most vulnerable people unable to understand or make sense of some aspects of their experiences. Alternatively, they understand only too well, but lack the channels to be heard. These injustices, in turn, might also lead to a situation where individuals question their right to exist and give up all claims in this respect.

Interpretive (hermeneutical) inequality is very difficult to detect and change, since privileged persons tend to understand those things that it serves them (us) to understand. Consequently, marginalised groups' interpretations of social phenomena remain silenced, even though they might very well be valid and even openly and explicitly requested.

This silencing is indeed a challenge for initiatives such as the Asylum Commission, which gathers actors with various power resources in an attempt to build solidary alliances for transformative knowledge production in the field of asylum rights. In a recent book about the asylum regime, Picozza (2021) reflects on these challenges through ethnographic fieldwork. She argues that the figure of the "refugee" produces its counterpart, the "refugee supporter", "as an embodiment of a specifically European and postcolonial 'good' whiteness, premised on liberal, democratic and humanitarian discourses" (Picozza 2021, xxiii).

Testimonial injustice is easier to spot than hermeneutical injustices. It occurs when a speaker receives less credibility as a result of negative identity prejudice. This injustice is perpetrated against members of groups whose testimonies are questioned and disbelieved because of negative prejudicial stereotypes about those groups. A typical example is when someone seeking protection is not deemed credible due to certain assigned identities, be it "woman", "child", or "HGBTQI+". Another example that has been evident in the Asylum Commission is when people seeking refuge in Sweden turn to politicians to point out shortcomings that have occurred in their asylum process. Such political actions are often based on an experience that the process has not been legally secure and that the concerned people have not been listened to. But for various reasons that may be related to limitations in society's processes for collective understanding and preconceived notions about people seeking asylum, it is very difficult to have one's voice heard. The difficulty is inflated by a strong belief in Sweden as a state governed by the rule of law. It becomes difficult, or even impossible, to put forward arguments that involve a legitimate critique. The image that the Swedish asylum procedure is legally secure seems to have such a deep anchoring in the migration administration's self-understanding that it is difficult to dislodge. This self-deception becomes particularly clear if we consider the "non-political" migration courts, where a ceremonial version of administrative justice is performed, making it extremely challenging to reveal injustices. Indeed, this also applies to the broader asylum regime, an order of inequalities (Achiume 2019) that is overshadowed by legalistic arguments and voluntary initiatives that aim to help (Picozza 2021).

What can we, as intellectual activists, do to contest these epistemological injustices? A way forward is offered by the emerging methodological approach referred to as "a scholarship of hope".

HOPE

In a recent anthology developed within the Asylum Commission framework (Elsrud et al. 2021), gender scholar Diana Mulinari describes hope as a concept that captures the vulnerability of life but also a human capacity to act and create other worlds. In this context, the organisation in safe spaces (often behind closed doors), various support groups, legal advice, fundraising, etc., may be understood as a practice of intellectual action and a potential methodological approach that may fruitfully articulate academic knowledge with a hopeful political vision. In this context, slogans such as *No frontiers, No nations, Stop deportation; No one is illegal; Not in my name*, are a form of hope for solidarity with refugees.

While Fricker inspires intellectual activism, Mulinari and her colleagues (2018, 2020) provide helpful reflections about hope as a form of scholarship, and this is an essential aspect of conversations concerning "What should we do?" In particular, their review of the processes at the turn of the millennium, when intersectionality took root in Swedish academia, is truly relevant. Intersectional research approaches had an explicit ambition at the time, to be intellectually activistic and provide politically relevant analyses by conceptualising the effect of economies' interdependence on the lives of human beings in a postcolonial world. Intersectionality also brought an ambition to explore and contest inequalities that are produced and reproduced in (white) academic work. The concept was born out of the black feminist critique of hegemonic "white feminism" and its knowledge production that lacked an understanding of the effects of racism. Moreover, intersectional perspectives were introduced to understand the dynamics of power at a time when the logic of capitalist accumulation was articulated through rationalising and effectivising social relationships.

Inch by inch, Mulinari and de los Reyes explain intersectionality as a perspective and political project has then endured a process of academisation and a dissociation of academic knowledge production from everyday struggles. It became "dislocated from the richness and heterogeneity of the Black radical tradition" (De los Reyes and Mulinari 2020, 188). Rather than politicising and re-politicising inequalities, endless discussions took place "about how to put categories together and what categories matter" (De los Reyes and Mulinari 2020, 188). This development went so far that intersectionality was eventually spoken of without naming the role of nation-states in the operations of power and categorisation practices.

One reason that fuelled the depoliticisation was that certain influential scholars such as Kimberlé Crenshaw and Patricia Hill Collins emerged as those who represented the intellectual field. This academisation went hand-in-hand with a gradually increasing inattention to the ground-up struggles that provided "the historical, political and epistemological space for the understanding of their theoretical work" (De los Reyes and Mulinari 2020, 188). As a result, knowledge also became depoliti-cised, allowing for an "unawareness" and a legitimised lack of interest in the conditions of black women's lived experiences as well as in the "heterogeneity of black women's intellectual production" (De los Reyes and Mulinari 2020, 191). It is against this background that Mulinari and her colleagues discuss transdisciplinary visionary research (see Martinsson and Mulinari 2018).

What can we learn from these reflections on translations of intersec-tionality? In retrospect, some of the epistemological injustices present today derive from the academisation of engaged and critical scholar-ships such as intersectionality. This circumstance is why scholars of hope should come back to reflect on "What should we do?" *and* simulta-neously engage practically in work that overcomes boundaries between academic work and political engagement. Put differently, to get engaged with: Transdisciplinary struggles to change the uncertain and insecure conditions that permeate the lives and existence of many human beings in contemporary times—"a theory on the flesh" (De los Reyes and Mulinari 2020, 191)—challenging the modern capitalist system, and creating futures outside those provided by hegemonic social relations, are necessary elements of a scholarship of hope.

As I read Mulinari and colleagues, the answer to "What should we do?" is a constant invention and reinvention of formative spaces. I under-stand this as an on-going process with no clear and ready answers. In these spaces, various positions can be articulated, and common futures can be envisioned. Future-oriented questions may help in the conversa-tions: What would a society beyond racism, patriarchy, heteronormativity, and capitalism look like, and how can we reach it? What forms of care or interdependency between human and non-human life would evolve, and how can we make this care flourish?

Intellectual Activism

As intellectual activists, we thus need to commit to people we meet *and* imagine, and we must struggle for a different world than the one in which we live. In my own community of belonging, this means, among other things, listening carefully to lived experiences of legal (un)certainty and that those who have legal knowledge share this when it is requested or needed. Instead of separating roles, we should constantly let the roles of activists, researchers, and professionals interact. I see separating the roles as a direct impediment to gaining a deeper understanding of legal (un)certainty, the (non)right to asylum and racist border practices. The interaction of roles also implies practical attempts to create coexistence without obscuring tensions, to make room for diversity, and to refuse to simplify. This is no easy task. As the Asylum Commission has experienced, joining forces in one struggle is extremely challenging due to the diversity of those seeking sanctuary in Sweden whose asylum claims have not been recognised, the involved professionals, and the scholars. Yet, this plurality does not mean that we cannot learn from others and—momentarily—also learn as equals.

Acknowledgements I would like to thank my activist intellectual friends, Torun Elsrud, Mehek Muftee, and Emma Söderman, for valuable comments on this text.

Bibliography

Achiume, E. Tendayi. 2019. "The Postcolonial Case for Rethinking Borders." *Dissent* 66 (3): 27–32. Project MUSE.

Boochani, B., et al. 2020. "Transnational Communities for Dismantling Detention: From Manus Island to the UK." *Community Psychology in Global Perspective* 6 (1): 108–128–128. http://eprints.whiterose.ac.uk/163891/1/Boochani-etal.pdf.

Bourdieu, Pierre. 1977. "The Economics of Linguistic Exchanges." *Social Science Information* 16 (6): 645–668. https://doi.org/10.1177/053901847701600601.

Bozalek, Vivienne, and Michalinos Zembylas. 2017. "Towards a Response-able Pedagogy Across Higher Education Institutions in Post-apartheid South Africa: An Ethico-political Analysis." *Education as Change* 21 (2): 62–85. http://hdl.handle.net/10566/3235.

Collins, Patricia H. 2013. "Truth-Telling and Intellectual Activism." *Contexts* 12 (1): 36–41.

De los Reyes, Paulina, and Diana Mulinari. 2020. "Hegemonic Feminism Revisited: On the Promises of Intersectionality in Times of the Precarisation of Life." *NORA—Nordic Journal of Feminist and Gender Research* 28 (3): 183–196.

Elsrud, Torun, Sabine Gruber, and Anna Lundberg. (2021) Asylkommissionen – en antirasistisk forskningspraktik [The Asylum Commission—An Anti-racist Research Practice]. In *Anti-rasistiskt socialt arbete i tider av skiftande välfärds- och migrationsregimer* [Anti-racist Social Work in Times of Shifting Welfare and Migration Regimes], edited by J. Johansson, Å. Söderkvist Forkby, and U. Wernesjö. Lund: Studentlitteratur.

Fernandes, Sujatha. 2017. *Curated Stories: The Uses and Misuses of Storytelling*. Oxford: Oxford University Press.

Ghosh, Bimal. 2000. "Toward a New International Regime for Orderly Movements of People." In *Managing Migration: Time for a New International Regime?*, edited by Bimal Ghosh, 6–261. Oxford: Oxford University Press.

Grosfoguel, Ramon, Laura Oso, and Anastasia Christou. 2015. "'Racism', Intersectionality and Migration Studies: Framing Some Theoretical Reflections." *Identities* 22 (6): 635–652.

Martinsson, Lena, and Diana Mulinari, eds. 2018. *Dreaming Global Change, Doing Local Feminisms*. London: Routledge.

Picozza, Fiorenza. 2021. *The Coloniality of Asylum: Mobility, Autonomy and Solidarity in the Wake of Europe's Refugee Crisis*. Lanham, MD: Rowman and Littlefield.

Spivak, Gayatri Chakravorty. 1988. "Can the Subaltern Speak?" In *Marxism and the Interpretation of Culture*, edited by C. Nelson and L. Grossberg, 271–313. Urbana, IL: University of Illinois Press.

Yuval-Davis, Nira. 2011. *The Politics of Belonging: Intersectional Contestations*. London: Sage.

INDEX

© The Editor(s) (if applicable) and The Author(s) 2022
M. Sandberg et al. (eds.), *Research Methodologies and Ethical Challenges in Digital Migration Studies*, Approaches to Social Inequality and Difference, https://doi.org/10.1007/978-3-030-81226-3